High School Physics:
Master It With Ease

(1) Introductory CURRENT ELECTRICITY

Sudhir K. Sood, Ph.D.

© Copyright 2020 by Sudhir K. Sood, Ph.D.
All rights reserved. No part of this book may be reproduced, stored in retrieval system or transmitted in any form or by any means whatsoever except as permitted by applicable copyright law.

Preface

This book has been written for all school students who get their first systematic exposure to the subject at high school level.

I firmly believe that at the formative stage of first exposure to a subject, the students should not be left with any gaps or doubts in their understanding. Accordingly, I have tried to furnish the book with many student-friendly features some of which are listed below.

1. In general, a main topic of a chapter introduces and discusses several new and different ideas. In this book, every main topic is carefully broken into subtopics such that each subtopic deals with a single well-defined idea. The discrete ideas of sub-topics are then explained one after another. Experienced teachers of physics intuitively follow such a step-by-step method in the class-rooms to help students understand the subject satisfactorily. The exposition is further supported by a number of properly labelled diagrams and tables.

2. The book contains a plethora of solved examples, each one of which is accompanied by a mention of the precise subtopic on which it is based.

3. Each chapter contains a large corpus of carefully designed exercises, which are further subdivided into 'Very Short', 'Short Answer' and 'Long Answer' type questions.

These exercises also contain a broad range of objective-type questions, which include 'Multiple Choice type (MCQs)', 'Fill in the Blanks type', 'True or False type', 'Matching type' and 'Quiz type' questions.

Ample answers, hints and solutions to these exercises are provided to help students achieve maximum elucidation of each topic.

4. In order to encourage 'learning by doing'; a number of activities are included in every chapter. Each activity can easily be performed by the students by using readily available material in their homes and school-labs.

5. A 'Question Bank' in the question- answer format is included at the end of the book. It contains a large collection of most such questions as find favor with the examiners. Students can use them for revision and consolidation of their studies as well as to learn how to answer different types of questions during their exams.

I hope that with all these useful features, the book gets converted into a self-contained textbook, which should benefit students both during self-study as well as for home-work and exam-work.

Any communication from you regarding deficiencies of the book and suggestions for improvement are welcome. Your help in pointing out errors that might have escaped attention despite effort to produce error free manuscript will be greatly appreciated. You can contact me at sudhbapc@gmail.com

About the Author

Sudhir K. Sood earned his Ph.D. degree in fundamental particle physics from University of Delhi. Subsequently, as research scientist and Professor of Physics at Universities in France, Canada and India, Dr. Sood has taught a number of courses both at introductory and advanced graduate level. He has lectured at international Physics conferences and authored numerous well-cited research papers that are published in reputed peer reviewed journals. More recently, for more than a decade, he has taught students in Delhi who wish to specialize in engineering, medicine and physical science courses.

Other Titles by the Author

1. High School Physics: Master It With Ease (1) Introductory Current Electricity
2. High School Physics: Master It With Ease (2) Introductory Electromagnetism
3. High School Physics: Master It With Ease (3) Introduction To Reflection and Refraction of Light, Mirrors and Lenses
4. A to …Z Class 10 CBSE/NCERT Physics (For Indian students)
5. AP Physics C Electricity and Magnetism, 2020 Edition: 100 Must-Know Questions in 1. Electrostatics 2. Conductors, Capacitors, Dielectrics With Answers and Explanations

Contents

Preface 3
Contents 5
INTRODUCTION 9
    Some important properties of electric charges 9
CHAPTER 1 ELECTRIC CURRENT AND ELECTRIC CIRCUIT 10
    Electric current consists of flow of electric charge 10
    Definition of electric current 10
    Electric current can flow in a closed and continuous path only 10
    What is an electric circuit? 10
    Direction of electric current 10
    Unit of Electric Current 11
    Measurement of Electric Current 11
        SOLVED EXAMPLES 12
        EXERCISES E 1.1 12
        Answers 14
        Hints/Solutions 14
        SUMMARY OF THE CHAPTER 15
CHAPTER 2 ELECTRIC POTENTIAL DIFFERENCE AND ELECTRIC POTENTIAL 16
    What causes the flow of electric charge? 16
    How is the flow of electric charge in a circuit maintained? 16
    The mode of action of an electric cell for generating and maintaining potential difference 17
    Definition of potential difference 17
    Unit of Potential Difference 17
    Measurement of Potential Difference 17
    Potential at a point 18
        SOLVED EXAMPLES 18
        EXERCISES E 2.1 19
        Answers 20
        Hints/Solutions 21
        SUMMARY OF THE CHAPTER 21
CHAPTER 3 CIRCUIT DIAGRAM 22
    What is a circuit diagram? 22

Symbols of electric components 22
      SUMMARY OF THE CHAPTER 23

CHAPTER 4 OHM'S LAW 24
What is Ohm's law? 24
ACTIVITY A 4.1 24
Resistance and its unit 25
The resistance of a conductor affects the flow of current through it 26
Physical reason for the resistance of a metallic conductor 26
Different components offer different resistances 27
      SUMMARY OF THE CHAPTER 27

CHAPTER 5 ON WHICH FACTORS DOES THE RESISTANCE OF A CONDUCTOR DEPEND? ..28
ACTIVITY A 5.1 28
Electric Resistivity 28
Unit of resistivity 29
Electric resistivity of some important substances 29
TABLE T 5.1 29
Uses of low resistivity materials 30
      SOLVED EXAMPLES 30
      EXERCISES E 5.1 32
        Answers 36
        Hints/Solutions 36
      SUMMARY OF THE CHAPTER 38

CHAPTER 6 RESISTANCE OF A COMBINATION OF RESISTORS 39
Combination of Resistances 39
Equivalent resistance of a combination of Resistances 39
Resistors in series 39
ACTIVITY A 6.1 39
Equivalent resistance of a series combination 40
      SOLVED EXAMPLES 42
      EXERCISES E 6.1 45
        Answers 47
        Hints/Solutions 47
Resistors in Parallel 48
ACTIVITY A 6.2 48

Equivalent resistance of a parallel combination 49

Advantages of a parallel circuit over a series circuit 51

    SOLVED EXAMPLES 52

    EXERCISES E 6.2 58

    Answers 61

    Hints/Solutions 62

    SUMMARY OF THE CHAPTER 64

CHAPTER 7 HEATING EFFECT OF ELECTRIC CURRENT 65

Conversion of electrical energy into other forms of energy 65

Heating effect of current 65

Expression for heat produced in a resistor-Joule's Law 65

Consequences of Joule's law of heating 66

Practical Applications of heating effect of current 66

Disadvantages of heating effect of curren 67

    SOLVED EXAMPLES 68

    EXERCISES E 7.1 69

    Answers 71

    Hints/Solutions 71

    SUMMARY OF THE CHAPTER 72

CHAPTER 8 ELECTRIC POWER AND COMMERCIAL UNIT OF ELECTRIC ENERGY 73

Definition of electric power 73

Unit of electric power 73

Commercial unit of energy-kilowatt hour 73

Rating of electric appliances 74

Choice of an appropriate fuse 74

    Solved Examples 75

    EXERCISES E 8.1 79

    Answers 81

    Hints/Solutions 81

    SUMMARY OF THE CHAPTER 82

QUESTION BANK 83-113

8

INTRODUCTION

Current electricity plays a vital role in all walks of our life. In our homes and places of work, we use it for lightening and for running various appliances. In factories and hospitals, various machines, instruments and gadgets depend on electricity for their functioning. Its use is so indispensable that, quite often, all these places make use of standby electric generators and inverters in case the supply from the mains gets interrupted. One of the important reasons why electricity finds applications everywhere is that its usage is very convenient. It becomes available to us simply by pressing a switch-even a child knows how to illuminate a dark room by switching on an electric bulb.

In the present book, we shall study some basic features of current electricity. Starting with what constitutes an electric current and how and under what conditions it flows in a conductor, we go on to study Ohm's law which relates the current through a conductor to the potential difference across its ends. Next, we study some properties of resistance offered by the conductor to the flow of current. We also learn about different ways of combining the resistances. We end the book with some discussion of the heating effect of current and the electric power.

Before proceeding further, let us list some important properties of electric charges that you might already be familiar with.

Some important properties of electric charges

- Electric charge is responsible for producing all electrical phenomena.
- Two types of electric charge exist in nature. They are called positive and negative charges.
- All matter is made of very small constituents called atoms. Each atom contains elementary particles known as electrons and protons that have equal amount of negative charge and positive charge respectively. Since the number of electrons is equal to that of protons, a piece of matter is ordinarily electrically neutral.
- When two bodies are rubbed together, some negatively charged electrons get transferred from one of them to the other. For example, when a glass rod is rubbed with silk cloth, electrons are transferred from glass rod to the silk cloth. As a result, silk cloth has excess number of electrons and starts behaving like a negatively charged object. At the same time, the glass rod, which is deficient in electrons starts behaving like a positively charged object.
- Two charged objects either attract or repel each other. Experiments show that like (or similar) charges repel whereas unlike (or opposite) charges attract each other. The force between the charges is governed by the Coulomb's law.
- Experiments also show that the total electric charge of an isolated system remains constant with time. In other words, charge can neither be created nor destroyed. This is known as the law of conservation of charge and is considered as exact law of nature.
- There are some materials (such as metals and their alloys), in which electric charges can move quite easily. Such materials are called conductors.
- Those substances (for example, glass, rubber, dry paper and wood), in which electric charges cannot move freely are called insulators.
- When electric charges move, we say that electric current is flowing.
- The subject of electricity is divided into two parts. Static electricity is the study of physical phenomenon produced by charges at rest whereas current electricity is the study of physical phenomenon related to the electric charges in motion.

CHAPTER 1

ELECTRIC CURRENT AND ELECTRIC CIRCUIT

Electric current consists of flow of electric charge

Just as the water flowing in a river constitutes a current of water, *the flow of electric charge in a conducting path constitutes an electric current.*

Definition of electric current

Electric current through a cross-section of a conductor is defined as the net amount of electric charge flowing per unit time across it.

Let Q be the net charge flowing across the cross-section of the conductor in time t, then the current I is given as

$I = Q/t$ (1.1)

Electric current can flow in a closed and continuous path only

We are all familiar with the working of an electric torch. Figure 1.1 shows an arrangement similar to that of a torch. Here, an electric cell is joined to a bulb through a plug key (or a switch) with the help of thick conducting wire. The electric cell causes and maintains the flow of an electric charge through conducting wire and the bulb.

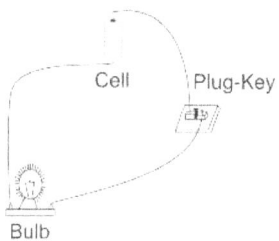

FIGURE 1.1 Schematic view of battery torch - an example of electric circuit

It is only when the key is plugged in (or the switch is on) that the bulb glows. In this case, there exists a closed and continuous path containing (i) an electric cell (or a battery, which is a combination of two or more cells), (ii) the conducting wires and (iii) the bulb. Electric charge flows through such a closed path. It is the electric charge circulating in the bulb, which is responsible for the latter's glowing. No wonder that the bulb stops glowing as soon as the closed and continuous path is broken, which happens when the switch is turned off.

This simple example shows that a closed and continuous path is essential for the flow of electric current.

What is an electric circuit?

An electric circuit is the closed conducting path in which an electric current can flow.

Direction of electric current

In case of metallic conductors, some of the negatively charged electrons can move freely from one atom to another. It is these free electrons, which comprise the flow of electric charges in conductors. Historically speaking, the discovery of electrons was made much later than the study of electric current. In earlier studies, the direction of current was taken as the direction of flow of positive charge. We continue to follow the earlier convention and, therefore, *the direction of current is taken as the direction of flow of positive charge or opposite that of the flow of electrons.*

Unit of Electric Current

SI unit of electric current is known as ampere and is denoted as A. It is named after the French scientist Andre-Marie Ampere (1775-1836). Eq. (1.1) allows us the way to define one ampere. In SI system of units, the unit of charge is coulomb (C) and the unit of time is second (s). So the SI unit of current is

1 ampere = 1 coulomb /1 second

$$1 A = 1 C s^{-1} \qquad (1.2)$$

We can, therefore, define that ***one ampere is that much current which flows through a cross-section of a conductor when one coulomb of charge flows through it in one second.***

For practical purposes, we make use of smaller units: 1 milliampere (mA) = 10^{-3} A or 1 microampere (μA) = 10^{-6} A.

We know that an electron and a proton have equal amount of negative charge and positive charge respectively. It is worth noting that an electron carries a negative charge of magnitude 1.6×10^{-19} C and a proton has an equal amount of positive charge.

Measurement of Electric Current

The electric current flowing in a circuit is measured by using an instrument called ammeter shown in Figure 1.2.

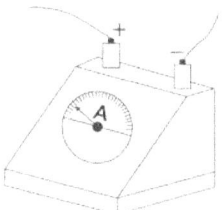

FIGURE 1.2 *Ammeter- the instrument used to measure electric current*

It has a needle which moves over a scale. The scale is so calibrated to as to directly give the measured current in amperes or some fraction of ampere. The ammeter has two terminals marked positive (+) and negative (-). Its (+) terminal should be joined to the positive terminal of the battery and (-) to the negative terminal of the battery.

An ammeter is connected in series with the circuit through which the current to be measured flows. It means that the ammeter is connected in such a way that whole of the current to be measured flows through it. Figure 1.3 shows a simple electric circuit containing an ammeter joined in series with a cell, a plug key and a bulb.

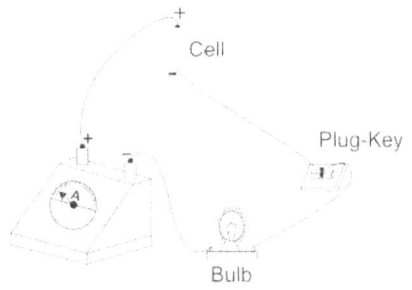

FIGURE 1.3 *The ammeter is joined in series with a cell, a plug key and a bulb.*

When the key is inserted, the circuit is closed and the bulb glows. The current flows in the circuit from the positive terminal of the cell to its negative terminal via the ammeter, the bulb and the plug-key. The value of current can be read on the ammeter-scale.

SOLVED EXAMPLES

1.1 Electric Charge and Current

If a current of 1.2 A flows through the filament of an electric bulb for 10 minutes, how much electric charge flows through the filament?

Solution

Current, $I = 1.2$ A

Time, $t = 10$ min $= 600$ s

The formula $I = Q/t$ gives

charge $Q = I \times t = (1.2 \text{ A}) \times (600 \text{ s}) = 720$ C

1.2 Electric Charge and Number of Electrons

Calculate the number of electrons required to get two coulombs of negative electric charge?

Solution

Total amount of negative charge = 2 C

The magnitude of negative charge of 1 electron = 1.6×10^{-19} C

Let the number of electrons in two coulombs of charge = n

Then, $n \times$ charge of 1 electron = total charge

So, n = total charge/ charge of 1 electron

$= 2 \text{ C} / (1.6 \times 10^{-19} \text{ C})$

$= 1.25 \times 10^{19}$.

1.3 Electric Charge and current

(Based on Higher Order Thinking Skills (HOTS))

Suppose 2.5×10^{19} electrons pass through a cross-section of a conducting wire in 15 seconds. What is the current flowing through the wire?

Solution

Magnitude of charge of one electron, $e = 1.6 \times 10^{-19}$ C

No of electrons passing through cross section, $N = 2.5 \times 10^{19}$

Total charge passing through cross section, $Q = N \times e$

$= (1.6 \times 10^{-19} \text{ C}) \times (2.5 \times 10^{19}) = 4.0$ C

Time, $t = 15$ s

So, current, $I = Q/t = 4.0 \text{ C} / 15 \text{ s} = 0.267$ A

EXERCISES E 1.1

Based on Electric Current and Circuit

A. Very Short Answer Type Questions

1. Name the particles whose flow constitutes the current in a metallic wire?
2. What is an electric circuit?
3. Define ampere.
4. Name the instrument used to measure electric current.
5. In an electric circuit containing a cell and a conductor, current flows through the conductor from which terminal of cell to which terminal?
6. How is the charge Q flowing across any cross-section of a conductor in time t related to the current I through it?
7. How much electric charge does an electron possess? What is its sign?
8. Electrons move from left to right in a conductor. In which direction does electric current flow?

B. Short Answer Type Questions – I

9. What is an ammeter? How it connected in an electric circuit?
10. What is the direction of conventional current? How is it related to the direction of flow of electrons? Why?
11. How much electric charge does an electron possess? What is its sign?
12. What are SI units of electric charge and electric current? How are they related?
13. *(Numerical Problem)* How much is the charge flowing through a copper wire in half an hour if it carries a current of 1.5 A?
14. *(Numerical Problem)* An electric bulb draws a current of 0.2 A when the voltage is 220 volts. Calculate the amount of electric charge flowing through it in one hour.
15. *(Numerical Problem)* Find the current flowing in a conductor in which 200 C of charge flows in 5 minutes.
16. *(Numerical Problem)* If a current of 2.0 A flows through a cross-section of area of 1 cm^2 for 10 minutes, how much electric charge flows through the area?

C. Short Answer Type Questions – II

17. Name an instrument that measures electric current in a circuit. Define the unit of electric current.
18. *(Numerical Problem Based on Higher Order Thinking Skills (HOTS))* Suppose 5.0×10^{20} electrons pass through a cross-section of a conducting wire in 10 seconds. What is the current flowing through the wire?
19. *(Numerical Problem)* How many electrons move through a conductor in one second if the current flowing through it is one ampere?
20. *(Numerical Problem Based on Higher Order Thinking Skills (HOTS))* How many electrons pass through a conductor per minute if a current of 5 A flows through it? (Given: magnitude of charge of an electron = 1.6×10^{-19} C).

D. Objective Questions

I. Multiple Choice Type Questions

Choose the Correct Answer:

21. A current of 4.8 A flows in a conductor. The number of electrons that cross any section per second is
(a) 10^{19}
(b) 2×10^{19}
(c) 3×10^{19}
(d) 3.3×10^{19}
22. The current which flows through a conductor when one coulomb of charge flows through any cross section in two seconds is
(a) 1
(b) 0.5 A
(c) 2 A
(d) none of the above
23. The instrument used to measure electric current flowing in a circuit is called

(a) ammeter
(b) cell
(c) voltmeter
(d) none of the above

24. One coulomb equals
(a) 1 ampere / 1 second
(b) 1 ampere × 1 second
(c) 1 volt × 1 second
(d) 1 volt / 1 second

II. Fill in the blanks Type

Fill in the blanks:

25. .... types of electric charge exist in nature.
26. A ......path is essential for the flow of electric current.
27. An ammeter is connected in …….with the circuit.

III. True or False Type

Mark the following Statements True (T) or False (F)

28. Current electricity is the study of physical phenomenon produced by charges at rest. T/F
29. When a glass rod is rubbed with silk cloth, electrons are transferred from glass rod to the silk cloth. T/F
30. The direction of current is taken as the direction of flow of electrons in metallic conductors. T/F
31. The electric current flowing in a circuit is measured by using an instrument called ammeter. T/F

Answers

1. Electrons.
4. Ammeter.
5. The current flows through the conductor from the positive terminal of the cell to its negative terminal.
8. Right to left.
11. Magnitude= 1.6×10^{-19} C. Sign: negative.
13. 2700 C
14. 720 C
15. 0.667 A.
16. 1200 C
18. 8.0 A.
19. 6.25×10^{18}.
20. 1.89×10^{21}.
21. (c)
22. (b)
23. (a)
24. (b)
25. Two
26. closed
27. series
28. F.
29. T.
30. F
31. T.

Hints/Solutions

13. $Q = I \times t = (1.5 \text{ A}) \times (30 \times 60 \text{ s}) = 2700$ C.
14. $Q = I \times t = (0.2 \text{ A}) \times (60 \times 60 \text{ s}) = 720$ C.
15. $I = Q / t = 200 \text{ C} / (5 \times 60 \text{ s}) = 0.667$ A.
16. Here, current, $I = 2.0$ A, time, $t = 10$ min $= 600$ s. Charge $Q = I \times t = (2.0 \text{ A}) \times (600 \text{ s}) = 1200$ C.
18. Here, number of electrons, $n = 5.0 \times 10^{20}$ charge on one electron $= 1.6 \times 10^{-19}$ C. Charge, $Q = n \times$ charge on one electron $= (5.0 \times 10^{20}) \times (1.6 \times 10^{-19} \text{ C}) = 80.0$ C, Current, $I = Q / t = 80.0 \text{ C} / 10 \text{ s} = 8.0$ A.

19. Here, charge on one electron = 1.6×10^{-19} C, current, $I = 1$ A. By definition of ampere, 1 C of charge flows per second. Charge, $Q = 1$ C, Number of electrons = $Q/$ (charge on one electron) = 1 C/ (1.6×10^{-19} C) = 6.25×10^{18}.

20. Here, charge on one electron = 1.6×10^{-19} C, current, I = 5A. By definition of ampere, 5C of charge flows per second. Charge per minute Q = 60×5 C = 300 C. Number of electrons = $Q/$ (charge on one electron) = 300 C/ (1.6×10^{-19} C) = 1.89×10^{21}.)

21. $n = Q/$(charge on one electron) = 4.8 C /(1.6×10^{-19} C) = 3×10^{19}.

22. Current, $I = Q / t = 1$ C / (2 s) = 0.5 A.

24. By definition, 1 ampere = 1 coulomb /1 second or 1 coulomb = 1 ampere × 1 second.

SUMMARY OF THE CHAPTER

- The flow of electric charge constitutes an electric current.
- The closed path in which a current flows is called an electric circuit.
- The direction of current is taken as opposite to the direction of flow of electrons in circuits containing metallic conductors.
- Electric current across a cross-section of a conductor is defined as the net amount of electric charge flowing per unit time across the cross-section.
- Let Q be the net charge flowing across the cross-section of the conductor in time t, then the current I is given as $I = Q/t$.
- SI unit of electric current is known as ampere and is denoted as A. One ampere is that much current which flows through a conductor when one coulomb of charge flows through any cross section in one second.
- The electric current is measured by using an instrument called ammeter.
- An ammeter is connected in series with the circuit through which the current to be measured flows.

CHAPTER 2

ELECTRIC POTENTIAL DIFFERENCE AND ELECTRIC POTENTIAL

What causes the flow of electric charge?

We know that flow of electric charge constitutes current. But charges cannot move by themselves in a conductor.

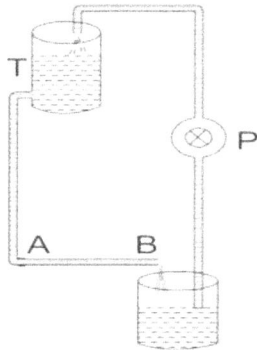

FIGURE 2.1 Water flows in a horizontal tube AB only when there is pressure difference between its two ends A and B.

In order to understand what causes the charges to move, let us take the analogy of water flow in a horizontal tube. In Figure 2.1, AB is a perfectly horizontal tube. Water cannot flow by itself from A to B in this tube. It starts flowing when the end A of the tube is connected to a tank T of water placed at a height so that end A is at higher pressure than end B. As a result, a pressure difference gets created between the two ends A and B of the tube. It is this pressure difference which is responsible for the flow of water in the horizontal tube.

An electric conductor is analogous to the horizontal tube. Just as water flows through the tube only when there is pressure difference between its ends, electric charge flows across a conductor only if there is a difference of electric pressure (called electric potential difference or simply potential difference) between its two ends.

It is the potential difference across the conductor that causes the flow of electric charges in the conductor. The potential difference can be produced by an electric cell (or a battery).

How is the flow of electric charge in a circuit maintained?

Again, consider the analogy of flow of water. If the continuous flow of water in the tube AB is to be maintained, we need a pump P, which does work on the water flowing out of tube at the end B and puts it back into the tank T (Figure 2.1). In other words, the pump takes the water from lower pressure to higher pressure, so that a pressure difference is continuously maintained between the two ends A and B of the tube.

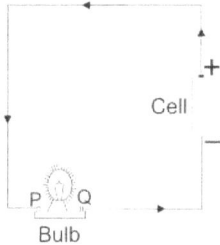

FIGURE 2.2 Electric current flows in the filament PQ of a bulb only when there is potential difference between its two ends P and Q.

Now consider Figure 2.2, which shows an electrical circuit containing a conductor PQ (the filament of a bulb) and an electric cell. In this case, it is the electric cell, which does work on the positive electric charges that enter it. By doing such work, the cell drives these charges from lower potential to higher potential, so that a potential difference is maintained between the two ends P and Q of the conductor and the current flows continuously in the conductor PQ.

The mode of action of an electric cell for generating and maintaining potential difference

An electric cell is essentially a device having two plates (called terminals) immersed in a chemical. Due to chemical reactions taking place within the cell, one of the terminals gets positively charged and is known as positive terminal while the other (called negative terminal) gets negatively charged. Now, a positively charged object is said to be at a higher potential than a negative charged object. The positive terminal, being a positively charged object is at a higher potential than a negative charged negative terminal. Thus, there develops a potential difference between the positive and the negative terminal of the cell. This happens even when the cell is not a part of the circuit.

Suppose we join the two terminals of the cell to the two ends of a conductor. Since similar charges repel each other, the negative terminal having negative charge repels the freely moving electrons in the conductor. They start moving along the conductor towards the positive terminal. Since, by convention, the direction of current is taken as opposite that of the flow of electrons, we find that the current flows in the conductor from positive to negative terminal. Also, the end of the conductor (joined to the positive terminal of the cell) attains a higher potential than the other end. We thus find that a potential difference develops between two ends of a conductor also and the current flows from higher to lower potential in it.

At the same time, the ongoing chemical action within the cell moves more positive charges towards positive terminal, which already has positive charge. Simultaneously, the same chemical action moves negative charges towards negative terminal, which already has negative charge. In this process, while trying to move positive (negative) charges towards positively (negatively) charged terminal, the cell has to do work against force of repulsion between like charges.

For doing such work and, hence, to maintain the potential difference, the cell makes use of the chemical energy stored within it.

Definition of potential difference

We have seen that an electric cell does work on the electric charge to maintain potential difference between two ends of the conductor. We make use of this idea to define potential difference between two points as follows. ***Electric potential difference between two points on a conductor carrying current is the work done to move a unit charge from one point to the other.***

Let W denote the work done for carrying a charge Q between the two points. Then the potential difference V between the two points is given by

Potential difference = work done / charge

or $V = W/Q$ (2.1)

Unit of Potential Difference

The SI unit of potential difference is called ***volt*** and is denoted by the symbol V. It is named after an Italian scientist A. Volta (1745-1827) who invented the electric cell. From Eq. (2.1), if W = 1 joule, Q = 1 coulomb, then 1 volt = 1 joule / 1coulomb, or

$1 V = 1 J C^{-1}$ (2.2)

Thus, ***one volt is the potential difference between two points when one joule of work is done to move a charge of one coulomb from one point to the other.***

Measurement of Potential Difference

The potential difference between two points is measured by using an instrument called voltmeter (Figure 2.3). Like an ammeter, a voltmeter also has a needle which moves over a scale. The scale is so calibrated to as to directly give the measured value of potential difference in volts or fraction of a volt. The voltmeter has two terminals marked positive (+) and negative (-). Its (+) terminal should be joined to the positive terminal of the battery and (-) to the negative terminal of the cell or the battery.

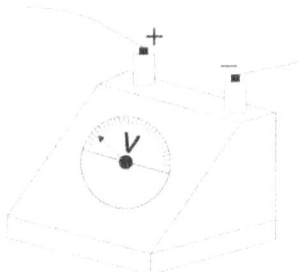

FIGURE 2.3 *Voltmeter - the instrument used to measure potential difference.*

Now, consider a circuit containing a bulb connected to a cell through a plug key. We can measure the potential difference across the glowing bulb only. This happens when the key is inserted so that the current flows in the circuit (as measured by an ammeter), which makes the bulb glow. We connect a voltmeter V *across* the bulb as shown in Figure 2.4. We say that *the voltmeter is joined in parallel across the bulb.* Note that in parallel connection, the same potential difference exists across the bulb and the voltmeter.

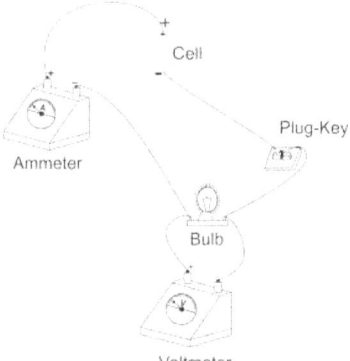

FIGURE 2.4 *In order to measure the potential difference across the glowing bulb, the voltmeter is joined in parallel across the bulb.*

Potential at a point

Although the physical quantity responsible for the flow of electric current between two points A and B in a conductor is the potential difference between A and B, yet, for completeness sake, we proceed to define the potential at a point as follows.

First, we choose our reference point, where the potential is taken to be zero. Then the potential of any other point P is defined as the potential difference between P and the reference point. Conventionally, the reference point is chosen to be infinity. Consequently, we define *the potential at any point as the amount of work done in bringing a unit charge from infinity to that point.*

SOLVED EXAMPLES

2.1 Potential Difference and Work Done

A charge of 3 C moves across two points, which have a potential difference of 10 V between them. Calculate the amount of work done.

Solution

We have potential difference $V = 10$ V

Charge $Q = 3$ C

So, work done $W = V \times Q = 10$ V $\times$ 3 C = 30 J.

2.2 Potential Difference and Work Done
(Based on Higher Order Thinking Skills (HOTS))

A 10 V battery is supplying current to a circuit. How much energy is supplied by the battery to each coulomb of electric charge passing through it?

Solution

Charge $Q = 1$ C, Potential difference $V = 10$ V

Work done to move the charge through the battery, $W = V \times Q = 10$ V $\times$ 1 C = 10 J

An amount of energy equal to the work done must be given to the charge

So, the energy given to each coulomb of charge = 10 J.

EXERCISES E 2.1

Based on Electric Potential and Potential Difference

A. Very Short Answer Type Questions

1. Name the SI unit of potential difference.
2. Define electric potential difference between two points?
3. How is the potential difference between two points related to the work done and the charge moved from one point to the other?
4. How is the potential at a point defined?

B. Short Answer Type Questions – I

5. Which instrument is used to measure the potential difference between two points? How is it connected in the circuit?
6. Name a device which can maintain a potential difference across a conductor. What is the source of energy of this device?
7. What causes the flow of electric charges along a conductor? How can such a flow be maintained?
8. What is the SI unit of potential difference? How is it defined?
9. *(Numerical Problem)* Calculate the potential difference between the two terminals of a battery if 40 J of work is done in transferring 4 C of charge from one terminal of the battery to the other.
10. *(Numerical Problem)* What is the potential at a point P if 2 J of work is done in bringing a particle having an electric charge of 0.5 C from infinity to P?
11. *(Numerical Problem)* How much energy is given to 5 C of charge passing through a 12 V battery?
12. *(Numerical Problem)* How much charge is passing between two points having a potential difference of 6 V if the amount of work done on the charge is 3 J?

C. Short Answer Type Questions – II

13. *(Numerical Problem Based on Higher Order Thinking Skills (HOTS))* What is the work done when one million electrons in a conductor move from a point P at a potential of 100 V to another point Q at a potential of 200 V? (Given: magnitude of charge of one electron = 1.6×10^{-19} C).

D. Objective Questions

I. Multiple Choice Type Questions

Choose the Correct Answer:

14. 1 joule equals
(a) 1 volt / 1 coulomb
(b) 1 volt × 1 coulomb
(c) 1 coulomb / 1 volt
(d) 1 ampere / 1 coulomb

15. The instrument used to measure electric potential difference between two points in a circuit is called
(a) voltmeter
(b) cell
(c) ammeter
(d) watt meter

16. The amount of work done when a charge of 6 C moves across two points, which have a potential difference of 2 V between them, is
(a) 3 J
(b) (2 /6) J
(c) 12 J
(d) none of the above

17. *(Based on Higher Order Thinking Skills (HOTS))* A point P in a conductor is at a potential of 10 V and another point Q is at a potential of 10 V
(a) electrons will move from point P to point Q
(b) electrons will move from point Q to point P
(c) no electron will move from point P to point Q
(d) none of the above

18. In order to measure electric potential difference between two points P and Q, we make use of a voltmeter. Which of the following statement is correct?
(a) we connect a voltmeter V in series with P and Q
(b) the current flowing through the voltmeter is different from that between P and Q
(c) same current flows through the voltmeter as between P and Q
(d) none of the above

II. Fill in the blanks Type

Fill in the blanks:

19. It is the ………across the conductor that causes the flow of electric charges in the conductor.
20. The cell uses the …… energy stored within it to maintain the potential difference.
21. 1 volt = 1 ….. / 1 coulomb.
22. The .... is joined in .... across the bulb.
23. In the symbol used to represent a cell, the longer line represents the ....terminal while the smaller line represents the .... terminal of the cell.

III. True or False Type

Mark the following Statements True (T) or False (F)

24. *(Based on Higher Order Thinking Skills (HOTS))* One of the terminals of an electric cell is at a higher potential than the other only when the cell is a part of a circuit carrying current. T/F
25. One volt is the potential difference between two points when one joule of work is done to move any charge from one point to the other. T/F
26. The positive terminal of a voltmeter is joined to the positive terminal of the cell or the battery. T/F

Answers

9. 10 V.	**15.** (a)	**21.** joule
10. 4 V.	**16.** (c)	**22.** voltmeter, parallel
11. 60 J.	**17.** (c)	**23.** positive, negative
12. 0.5 C	**18.** (b)	**24.** F
13. 1.6×10^{-11} J.	**19.** potential difference	**25.** F
14. (b)	**20.** chemical	**26.** T

Hints/Solutions

9. $V = W/Q = 40$ J/4 C $= 10$ V.

10. Potential at P is $V = W/Q = 2$ J/0.5 C $= 4$ V.

11. Energy given = work done. $W = VQ = (12 \text{ V}) \times (5 \text{ C}) = 60$ J.

12. $Q = W/V = 3$ J/6 V $= 0.5$ C.

13. Potential difference between P and Q = 200 V - 100 V = 100 V.
Charge flowing $Q = n \times$ *charge of one electron* $= 10^6 \times 1.6 \times 10^{-19}$ C $= 1.6 \times 10^{-13}$ C. Work done, $W = VQ = (100 \text{ V}) \times (1.6 \times 10^{-13}$ C$) = 1.6 \times 10^{-11}$ J.

14. By definition, 1 volt = 1 joule/1coulomb. or 1 joule = 1 volt × 1 coulomb.

15. The potential difference between two points is measured by using an instrument called voltmeter.

16. Work done $W = VQ = 2$ V $\times 6$ C $= 12$ J.

17. It is the potential difference between two points that causes the flow of electric charges between them. Here, potential difference = 0 V.

18. It is the potential difference, which is same across voltmeter and between two points P and Q but the current flowing through the voltmeter is different from that between P and Q.

SUMMARY OF THE CHAPTER

- The potential difference across the conductor causes the flow of electric current in the conductor.
- An electric cell or a battery (i.e. combination of cells) maintains the potential difference needed for the flow of a steady current in the circuit.
- Electric potential difference between two points on a conductor carrying current is the work done to move a unit charge from one point to the other or $V = W/Q$.
- The SI unit of potential difference is called volt
- One volt is the potential difference between two points when one joule of work is done to move a charge of one coulomb from one point to the other.
- The potential at any point is the amount of work done in bringing a unit positive charge from infinity to that point.
- The potential difference between two points is measured by using an instrument called voltmeter.

CHAPTER 3

CIRCUIT DIAGRAM

What is a circuit diagram?

A schematic diagram which shows the actual arrangement of various electric components in an electric circuit is called a circuit diagram.

Symbols of electric components

It is convenient to draw a circuit diagram by making use of symbols to represent different components. In Figure 3.1, we give the symbols of a few components that are used quite frequently in current electricity.

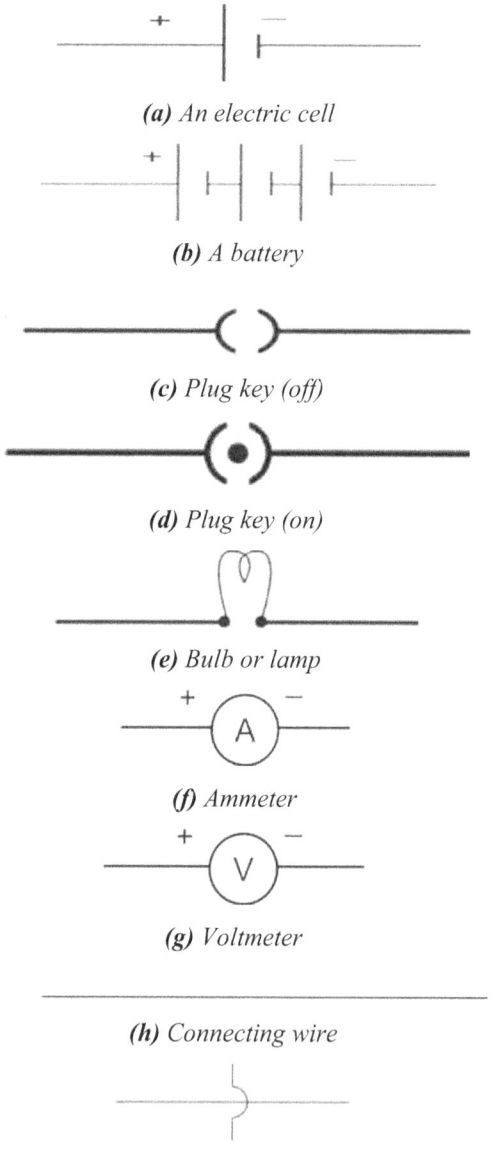

(a) An electric cell

(b) A battery

(c) Plug key (off)

(d) Plug key (on)

(e) Bulb or lamp

(f) Ammeter

(g) Voltmeter

(h) Connecting wire

(i) One connecting wire crossing anther without making connection

FIGURE 3.1 *Symbols of a few frequently used components*

Note that a cell is represented by two lines of unequal length. The longer line represents the positive terminal while the smaller line represents the negative terminal of the cell. A battery, which is a combination of cells, is obtained by successively joining the negative terminal of one cell to the positive terminal of the adjacent cell.

To illustrate the use of such symbols, we have drawn in Figure 3.2 a circuit diagram of the circuit of Figure 2.4 (of chapter 2) containing a cell, a closed plug key K, an electric bulb and an ammeter. Connecting wires are being used to join various components to each other and a voltmeter is joined across the lamp.

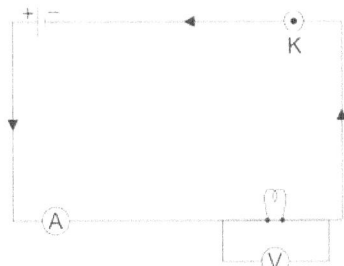

FIGURE 3.2 *A circuit diagram of the circuit of Figure 2.4 - drawn by using symbols of a cell, an ammeter, an electric bulb with a voltmeter across it and a closed plug key K. The arrows show the direction of current in the closed circuit.*

A comparison of these two figures shows that it is much easier to represent electric circuits by making use of symbols in place of drawings of actual components.

SUMMARY OF THE CHAPTER

- A schematic diagram which shows the actual arrangement of various electric components in an electric circuit is called a circuit diagram.
- It is convenient to draw a circuit diagram by making use of symbols to represent different components.

CHAPTER 4

OHM'S LAW

What is Ohm's law?

Ohm's law is one of the most widely used laws in current electricity. It was discovered by a German physicist G. S. Ohm (1787-1854) in 1827.

Ohm's law gives a relation between the current flowing through a conductor (say a metallic wire) and the potential difference across its ends. It states that ***the current flowing through a conductor is directly proportional to the potential difference across its two ends provided the temperature remains constant.***

Let I be the current flowing through the conductor and V be the potential difference across its ends, then, according to Ohm's law

V is proportional to I

or V = constant × I

or V / I = constant

The constant of proportionality is known as the *resistance* of the conductor and is denoted by the letter R. So,

$V/I = R$ *(4.1(a))*

or $V = R\,I$ *(4.1(b))*

Let us perform activity A 4.1 to verify Ohm's law for ourselves.

ACTIVITY A 4.1

The aim of this activity is to study the relationship (Ohm's law) between the potential differences across a conductor and the current through it.

The apparatus includes a nichrome wire of about half a meter length, four dry cells of about 1.5 V each, a voltmeter, an ammeter, a key and some thick copper wire for making connection.

The procedure consists of the following steps:

- Draw the circuit diagram as shown in Figure 4.1. Here R denotes the nichrome wire.

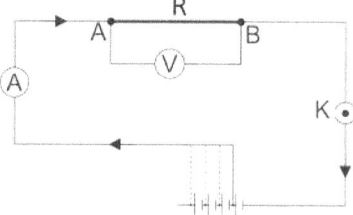

FIGURE 4.1 *The electric circuit to study Ohm's law*

- Arrange the various components as per the circuit diagram. The wire R is joined between points A and B. Use thick copper wire for joining the components to one another.
- First, use only one cell in the circuit. Insert the plug key

- Note the readings of the ammeter and the voltmeter V. The ammeter reading gives the value of the current I through the circuit including the nichrome wire. The voltmeter reading gives the potential difference V across the two ends of the wire. List these values in the form of the Table T 4.1 shown below.

TABLE T 4.1

S. No.	Number of cells	I	V	(V/I)
1	1			
2	2			
3	3			
4	4			

- Remove the key and connect two cells in series with the circuit. Reinsert the key and note the readings of the current I in the ammeter and the potential difference V in the voltmeter. List the new values in the table.
- Repeat the experiment by bringing in three cells and finally, all the four cells in the circuit, measuring and tabulating the values of I and V each time.
- Each time calculate the ratio of V to I.
- Plot the graph of the potential difference V against the current valves I.

An important precaution that needs to be observed is as follows.

The key should be inserted into the plug only when taking the readings of voltmeter and ammeter and should be removed after these measurements. Otherwise the current flowing in the circuit will cause heating of the resistors. This may not only adversely change the values but the cells may also get exhausted.

What we observe is as follows.
(i) The ratio V/I is found to be almost equal in various steps of the experiment.
(ii) The V-I graph is found to be a straight line (Figure 4.2) passing through the origin O.

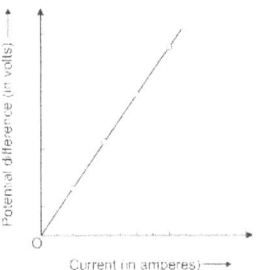

FIGURE 4.2 The graph between the potential difference and the current is a straight line, which is consistent with Ohm's law

These results are consistent with Ohm's law, according to which potential difference V is proportional to the current I, or, the ratio V/I is a constant.

The constant ratio V/I gives us the value of the resistance of the nichrome wire. The resistance R has a constant value for a given conductor at a given temperature.

Resistance and its unit

From Eq. (4.1 (a)), we have

$R = V/I$

So, *resistance of a conductor may be defined as the ratio of the potential difference across the conductor to the current flowing through it.*

The SI unit of resistance is called ohm and is denoted by the Greek letter Ω. If $I = 1$ A, $V = 1$ V, then from above equation, we have,

1 ohm = 1volt / 1ampere, or

$1 \Omega = 1 \, V \, A^{-1}$ (4.2)

Thus, one ohm is the resistance of a conductor having a current of one ampere flowing through it when potential difference across its ends is one volt.

A conductor having noticeable value of resistance is called a resistor.

The resistance of a conductor affects the flow of current through it

To see it, let us rewrite Ohm's law (Eq. 4.1 (a) and 4.1 (b)) as

$I = V/R$

The above equation tells us that for a constant potential difference V, current I is proportional to $(1/R)$, i.e. the current is inversely proportional to resistance. If, for example, the resistance in a circuit is doubled, the current reduces to half of its earlier value.

We can make use of the above property to vary the current in the circuit with the help of a variable resistance instead of changing the potential difference of the cell. *Such **a component having variable resistance, which is used to regulate the current in a circuit, is known as a rheostat.***

The symbols of both an ordinary resistor and a variable resistor or a rheostat are given in Figure 4.3.

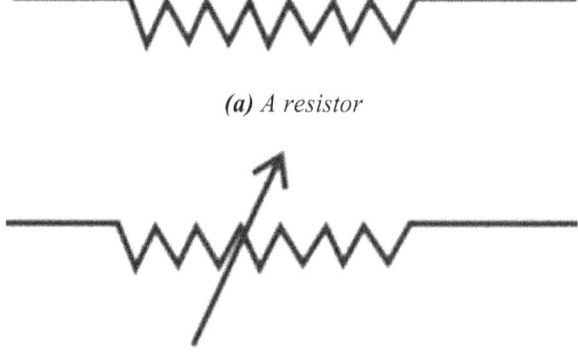

(a) A resistor

(b) A variable resistor or a rheostat

FIGURE 4.3

Physical reason for the resistance of a metallic conductor

In absence of applied potential difference, the free electrons within a conductor move randomly in all directions. However, when potential difference is applied, they start moving (or drifting) in one direction – thus constituting electric current. We know that the direction of flow of current is opposite to that of the drift of electrons.

The drifting electrons are not totally free to move because they collide with much bigger sized atoms present within the body of the conductor (Figure 4.4). They experience the force of attraction due to these atoms. Consequently, their motion gets retarded. Such retardation is similar to that experienced by water flowing in a pipe due to friction. *The resistance of a metallic conductor is a manifestation of friction-like effect experienced by drifting electrons in the conductor due to force of attraction of atoms of the conductor.*

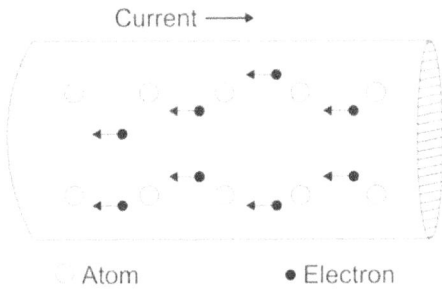

FIGURE 4.4 *The resistance of a metallic conductor is due to friction-like force experienced by drifting electrons in the conductor, which is similar to that experienced by water flowing in a pipe.*

Different components offer different resistances

While dealing with electric circuits, we come across various components (such as wires made of metals and alloys or filaments of bulbs of various types), which offer resistance to the flow of current. All these components differ in their resistance values.

A component which offers low resistance is a good conductor. Such a component provides an easy path to the flow of current.

On the other hand, *a component with high resistance opposes the flow of current and acts as a poor conductor.*

A component with very high resistance behaves like an insulator. Such a component practically stops the flow of current through it.

SUMMARY OF THE CHAPTER

- Ohm's law states that the current flowing through a metallic wire is directly proportional to the potential difference across its two ends provided its temperature remains constant
- V being proportional to I means $V = R\,I$. Here R is a constant of proportionality called the resistance of the conductor.
- Resistance of a conductor may be defined as the ratio of the potential difference across the conductor to the current flowing through it.
- The SI unit of resistance is called ohm, which may be defined as the resistance of a conductor having a current of one ampere flowing through it when potential difference across its ends is one volt.
- For a constant potential difference V, current I proportional to $1/R$, i.e. the current is inversely proportional to resistance
- The resistance of a metallic conductor is a manifestation of friction-like effect experienced by drifting electrons in the conductor due to force of attraction of atoms of the conductor.

CHAPTER 5

ON WHICH FACTORS DOES THE RESISTANCE OF A CONDUCTOR DEPEND?

In order to learn about the various factors affecting the resistance of a given conductor, we can perform the following activity.

ACTIVITY A 5.1

The aim of this activity is to learn about the factors on which the resistance of a conductor depends.

The apparatus includes a battery of four dry cells of voltage 1.5 V each, an ammeter, four conductors in the shape of uniform wires: (i) wire W_1, made of nichrome having certain length l and area of cross-section A, (ii) wire W_2, made of nichrome having same thickness as W_1 but twice the length of W_1, (iii) wire W_3, again made of nichrome having same length as W_1 but thicker than W_1 (iv) wire W_4 made of copper but with same length and cross-section as W_1, a plug key and some thick insulated copper wire for making the connection.

The procedure consists of the following steps:

- Draw the circuit diagram as shown Figure 5.1. Connect the various components using connecting wire leaving a gap PQ in the circuit.

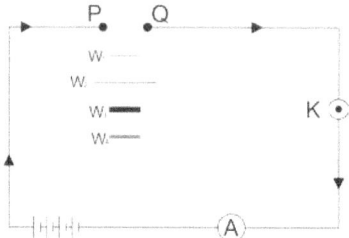

FIGURE 5.1 *The electric circuit to study the factors on which the resistance of a conductor depends*

- Connect the wire W_1 in the gap PQ.
- Insert the plug in the key. Note the reading of the ammeter. Let the current reading be I_1.
- Next, replace W_1 with W_2 and again note the current I_2 in the ammeter.
- In the same manner, replace W_2 with W_3 and note the value I_3 of the current.
- Finally join the copper wire W_4 in the circuit. Note down the value of current I_4 in the circuit.

Precaution The key should be inserted into the plug only when taking the readings of voltmeter and ammeter and should be removed after these measurements.

What we observe is as follows.

(i) The value of potential difference of (V) across all the wires remains the same.
(ii) The value of current I_2 is half that of I_1.
(iii) The value I_3 is larger than that of I_1.
(iv) I_4 is different from I_1, which means that the resistance of W_4 (made of copper) is different from that of W_1.

Combining various results and recalling that the current flowing through a conductor is inversely proportional to its resistance, we conclude that resistance of a conductor increases with increase of its length but decreases with increase of its cross-section. Moreover, it depends on the nature of the material of the conductor.

Electric Resistivity

Consider a conductor of given material having uniform thickness. Detailed experiments tell us that the resistance of the conductor is

(i) directly proportional to its length (l)

(ii) inversely proportional to its area of cross-section (A).

Combining the above two results, we have,

R is proportional to (l/A)

or *R = constant × (l/A)*

The constant of proportionality is called the *resistivity* of the material of the conductor and is denoted by ρ (Greek letter 'rho'). So,

$R = \rho \, l / A$ (5.1)

In Eq. (5.1), let $A = 1$ unit, $l = 1$ unit, then $\rho = R$, and therefore, *resistivity of a material is numerically equal to the resistance per unit length of a uniform conductor of that material having unit area of cross-section.*

Resistivity is an important property of the material of the conductor. Its value is independent of physical dimensions of the conductor but varies with temperature.

Unit of resistivity

By rewriting Eq. (5.1) as $\rho = R \, A/l$ and substituting the SI units of resistance R as ohm, area A as (meter)^2 and length l as meter respectively, we find that the SI unit of resistivity is *ohm meter (or Ω m).*

Electric resistivity of some important substances

The resistivity values of some important substances are given in Table T 5.1. The resistivity of a substance varies with temperature. The listed values in the table correspond to a temperature of 20^0 C.

TABLE T 5.1

Metals	Silver	1.60×10^{-8}
	Copper	1.62×10^{-8}
	Aluminium	2.63×10^{-8}
	Tungsten	5.20×10^{-8}
	Nickel	6.84×10^{-8}
	Iron	10.0×10^{-8}
	Manganese	1.86×10^{-8}
	Chromium	12.9×10^{-8}
	Mercury	94.0×10^{-8}
Alloys	Manganin	44×10^{-6}
	Nichrome	100×10^{-6}
	Constantan	49×10^{-6}
Insulators	Glass	$10^{10} - 10^{14}$
	Rubber	$10^{13} - 10^{16}$
	Diamond	$10^{12} - 10^{13}$
	Dry paper	10^{12}

Careful study of this table tells us that, as regards resistivity, different materials can be classified into the following two types:

(a) Low resistivity materials

Metals and their alloys fall into this category. Whereas metals have resistivity values in the range of 10^{-8} Ω m to 10^{-6} Ω m, the alloys have higher values in the range of 10^{-5} Ω m to 10^{-4} Ω m. From Table T 5.1, we find that the resistivity of alloys like manganin (alloy of copper, manganese and nickel), nichrome (alloy of nickel, chromium, manganese and iron) and constantan (alloy of copper, and nickel) is higher than that of the component metals.

Both metals and their alloys are good conductors of electricity.

(b) High resistivity materials

Insulators like glass, rubber, diamond and dry paper fall in this category. Their resistivity is typically of the order of 10^{15} Ω m, which is much higher (about 10^{23} times) than that of metals and their alloys.

Uses of low resistivity materials

(a) Uses of metals

Copper and aluminium, being very good conductors, are used for electrical transmission lines. Tungsten is the preferred metal for making the filaments of electric bulbs since its melting point is very high.

(b) Uses of alloys

Wires made of alloys are generally used for making the heating elements of electric heating devices like electric iron, toaster and heater etc. We mention below the main reasons thereof.

(i) Alloys do not oxidise or burn easily.

(ii) Unlike pure metals, the resistance of alloys changes very little with temperature.

(iii) The resistivity of an alloy being higher than that of metals by a factor of about 25-100, we need a smaller length of wire made of an alloy than the one made of metal to have required value of resistance. This result follows from Eq. (5.1), which gives $l = R A /\rho$, i.e. l is inversely proportional to ρ, for given R and A.

SOLVED EXAMPLES

5.1 Application of Ohm's law

When a 6 V battery is connected across an unknown conducting wire, there is a current of 5.0 m A in the circuit. Find the value of resistance of the conducting wire.

Solution

Potential difference $V = 6$ V

Current $I = 5.0$ mA $= 5.0 \times 10^{-3}$ A

On applying Ohm's law,

Resistance $R = V/I = 6$ V$/ (5.0 \times 10^{-3}$ A$) = 1.2 \times 10^{3}$ Ω.

5.2 Application of Ohm's law

(Based on Higher Order Thinking Skills (HOTS))

A metallic wire joined across a battery of 12 V has a current of 2 A flowing through it. What will be the value of current if the same wire is joined across a battery of 36 V?

Solution

Potential difference $V = 12$ V

Current $I = 2$ A

According to Ohm's law, the current through a given metal wire is proportional to the potential difference across its ends. New potential difference $V' = 36$ V.

Since V becomes three times the earlier value, the current will also become three times its original value.

So, new value of current $I' = 3 \times (2\ A) = 6\ A$.

5.3 Calculation of resistivity

A conductor wire of length 2.0 m and diameter 4 mm has a resistance of 16 Ω at room temperature. Find its resistivity. In your opinion, is the wire is made of pure metal or alloy?

Solution

Length of wire, $l = 2.0$ m, diameter, $d = 4$ mm $= 4 \times 10^{-3}$ m.

So, the area of cross-section $A = \pi d^2 / 4$.

The resistance, $R = 16$ Ω.

If ρ be the resistivity of the material of the wire, we have

$R = \rho l / A$, which gives

$\rho = R A / l = R \pi d^2 / (4 l)$

Or $\rho = (16\ \Omega)(3.14)(4 \times 10^{-3}\text{m})^2 / (4 \times 2.0\ \text{m})$

$= 100.5 \times 10^{-6}$ Ω m.

The above value of ρ tells us that the wire is made of alloy.

5.4 Change of Resistance

(Based on Higher Order Thinking Skills (HOTS))

The resistance of a metal wire is 5 Ω. It is then doubled on itself by folding. What will the new resistance?

Solution

Resistance of the original wire, $R_1 = 5$ Ω.

Let length and area of cross-section of the original wire be l and A respectively.

When it is doubled on itself by folding, its length becomes half the earlier value. Since the volume (= length × area of cross-section) of the wire remains constant, its area of cross-section, after folding, becomes twice as much as the earlier value.

So, the length and area of cross-section of the folded wire will be $(l/2)$ and $(2A)$ respectively.

If ρ be the resistivity of material of the wire, then

$R_1 = \rho l / A$

Resistance of the folded wire, $R_2 = \rho (l/2) / (2A) = (1/4) \times (\rho l / A)$

So, $R_2 / R_1 = 1/4$, which gives $R_2 = R_1/4 = 1.25$ Ω.

Hence the resistance of the folded wire will be 1.25 Ω.

5.5 Verification of Ohm's law

(Based on Higher Order Thinking Skills (HOTS))

When a current of 0.6 A flows through a conductor, the potential difference across it is 12 V. What will be the potential difference across the same conductor if the current value changes to 0.25 A?

Solution

According to Ohm's law, V/I = constant = R, the resistance of the conductor.

or, $V_2/I_2 = V_1/I_1$

Here, $I_1 = 0.6$ A, $V_1 = 12$ V, $I_2 = 0.25$ A, $V_2 = ?$

So, $V_2 = (I_2 \times V_1)/I_1 = (0.25 \text{ A} \times 12 \text{ V})/ 0.6 \text{ A} = 5 \text{ V}$.

5.6 Resistance and resistivity

A metallic wire has a radius of 0.25 mm and a resistance of 5 Ω. What will be the length of this wire if resistivity of the material of wire is 3.2×10^{-8} Ω m?

Solution

Radius $r = 0.25$ mm $= 2.5 \times 10^{-4}$ m.

Required resistance $R = 5$ Ω

Resistivity of the material of the wire $\rho = 3.2 \times 10^{-8}$ Ω m

Length of wire $l = ?$

Now, the area of cross-section of the wire $A = \pi r^2$.

We have, $R = \rho l / A$, which gives

$l = R A / \rho = R \pi r^2 / \rho$

or, $l = (5 \text{ Ω}) \times (3.14) \times (2.5 \times 10^{-4} \text{ m})^2 / (3.2 \times 10^{-8} \text{ Ω m})$

$= 30.7$ m.

So, required length of wire $l = 30.7$ m.

EXERCISES E 5.1

Based on Circuit Diagram, Ohm's Law, Resistance and Resistivity

A. Very Short Answer Type Questions

1. State Ohm's law.
2. Name the physical quantity which is the ratio of potential difference across a conductor to the current flowing through it.
3. Which physical quantity has the SI unit V A^{-1}?
4. What is the nature of graph between the potential difference across a conductor and the current flowing through it?
5. Name a metal which offers less resistance to the flow of current than copper?
6. What is the symbol used to represent resistance?
7. What is SI unit of resistance?
8. How is resistivity defined?
9. What is the SI unit of resistivity?
10. A wire of uniform cross-section is stretched to four times its length? By what factor does the resistivity change?
11. Name a metal used for electric transmission lines.
12. Which metal is used for making the filaments of electric bulbs?
13. How are the quantities volt, ampere and ohm related?

B. Short Answer Type Questions – I

14. What is the difference between metals and insulators in terms of their resistivity values?
15. *(Numerical Problem)* The resistance of a metallic wire is 2.5 Ω. How much current will it draw if it is joined across a battery having a potential difference of 12.5 V between its terminals?

C. Short Answer Type Questions – II

16. Define resistance of a conductor. Name and define its SI unit.
17. *(Based on Higher Order Thinking Skills (HOTS))* Two wires, one of the alloy manganin and the other of copper have equal length and resistance. Which one these wires will be thicker?

18. *(Numerical Problem)* The resistance of a metallic wire of length 1.5 m is 0.3 Ω. If the radius of the wire is 0.2 mm, what is the value of resistivity of the metal?

19. *(Numerical Problem)* When a conductor is connected to a 12 V battery, 0.6 A of current flows through it. What will be current value if it is connected to a 4.5 V battery?

20. *(Numerical Problem Based on Higher Order Thinking Skills (HOTS))* We have two copper wires A and B having identical mass. The length of B is half that of A. Find the ratio of resistances of B and A.

21. *(Numerical Problem Based on Higher Order Thinking Skills (HOTS))* The resistance of an insulated copper wire is 15 Ω. It is doubled by folding on itself. What will be the resistance of the folded wire?

22. *(Numerical Problem Based on Higher Order Thinking Skills (HOTS))* The resistance of a copper wire is 6 Ω. What will be the resistance of another copper wire whose length and diameter are both double those of the first wire?

23. *(Numerical Problem Based on Higher Order Thinking Skills (HOTS))* A wire of uniform cross section having a resistance of 4 Ω is stretched to three times its original length. What will be the resistance of the wire in the new situation?

D. Long Answer Type Questions

24. State Ohm's law. Express it mathematically. What is resistance?

25. (a) Express Ohm's law by a mathematical formula.
(b) Describe the activity with the help of a circuit diagram to establish the relationship between current I through a conductor and potential difference V across its two ends. Present this relationship graphically.

26. *(Numerical Problem Based on Higher Order Thinking Skills (HOTS))* The resistance of a copper wire is 2.0 Ω. What will be the resistance of another copper wire of double the diameter of the given wire but having the same volume as the given wire?

27. *(Numerical Problem)* What will be the diameter of an aluminum wire if its length is 14 m and its resistance value is 7 Ω? (Resistivity of aluminum is 2.8×10^{-8} Ω m)

E. Objective Questions

I. Multiple Choice Type Questions

Choose the Correct Answer:

28. Resistivity of a conductor depends upon its
(a) area of cross section
(b) resistance
(c) length
(d) none of the above characteristics

29. Why do we prefer constantan wire for making the heating elements of electric heating devices?
(a) low resistivity
(b) its resistance is independent of temperature
(c) high resistivity
(d) high resistance.

30. The length and area of cross-section of a resistor are doubled, its resistance will be
(a) halved
(b) unchanged
(c) doubled
(d) quadrupled.

31. Ohm's law is valid when the temperature of the conductor is:
(a) varying.
(b) very high
(c) very low
(d) constant

32. A potential difference of 10 V is applied across a 4 Ω resistance. The current in the conductor is:
(a) 40 A
(b) 2.5 A
(c) 2.4 A
(d) none of the above.

33. A wire is doubled by folding on itself. Its resistivity will be:
(a) halved
(b) unchanged
(c) doubled
(d) one fourth its earlier value.

II. Fill in the blanks Type Questions

Fill in the blanks:

34. The current flowing through a conductor is directly proportional to the ……… across its two ends provided the ……. remains constant.
35. ……of a conductor may be defined as the ……. of the potential difference across the conductor to the current flowing through it.
36. The resistance of a metallic conductor is a manifestation of …… -like effect experienced by drifting electrons in the conductor.
37. Resistivity of a material may be defined as the …… per unit length of a uniform conductor of that material that has a unit area of cross-section.
38. The SI unit of resistivity is ……..
39. *(Based on Higher Order Thinking Skills (HOTS))* The resistivity of an alloy is …… than that of metals by a factor of about …..

III. True or False Type Questions

Mark the following Statements True (T) or False (F)

40. One ohm is the resistance of a conductor having a current of one ampere flowing through it when potential difference across its ends is one volt. T/F
41. The resistance (R) of a uniform conductor is inversely proportional to its length. T/F
42. The resistance of a conductor varies with temperature. T/F
43. Pure metals have resistivity values in the range of 10^{-5} Ω m to 10^{-4} Ω m. T/F
44. Wires made of metals are generally used for making the heating elements of electric heating devices. T/F

IV. Matching Type Questions

45. Match the given symbols with their corresponding circuit elements

S. No.	Column A	Column B
1.	(a) An ammeter.	(i) —/\/\/\—
2.	(b) A closed plug key	(ii) —(v)—
3.	(c) A Cell	(iii) —◊—
4.	(d) Wires crossing without connection	(iv) —wwww—
5.	(e) Variable resistance/rheostat	(v) |
6.	(f) A resistor	(vi) —(A)—
7.	(g) A voltmeter	(vii) —⟩
8.	(h) Open plug key	(viii) —◊—

46. Select the pairs, in the two columns, that match each other

S. No.	Column A	Column B
1.	(a) an electric cell	(i) higher value of resistivity than that of metals
2.	(b) voltmeter	(ii) Ohm's law.
3.	(c) Resistivity	(iii) generation of potential difference
4.	(d) Ω m	(iv) measurement of the potential
5.	(e) The V-I graph is a straight line	(v) SI unit of resistivity
6.	(f) an alloy	(vi) value is independent of size of the conductor

V. Crossword Puzzle

47. Complete the crossword puzzle with the help of given clues

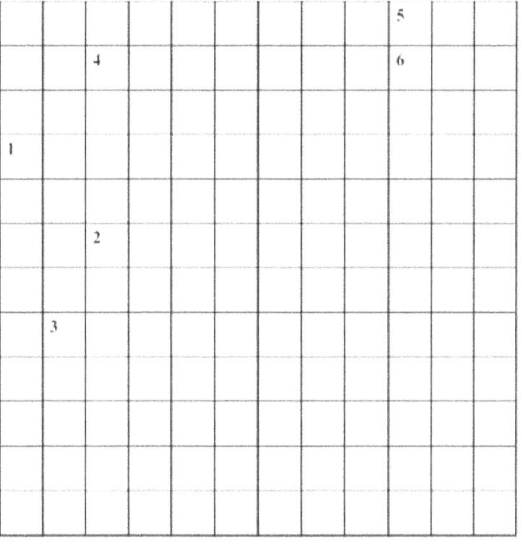

The Clues
Across:
1. An important property of the material of the conductor whose value is independent of physical dimensions of the conductor but varies with temperature (11 letters)
2. The type of combination of resistances in which the resistances are connected end to end so that the same current passes through each of them (6 letters)
3. The type of combination of resistances in which the resistances are joined together between two points such that a common potential difference exists across each of them (8 letters)
6. The SI unit of resistance (3 letters)

Down:
4. The ratio of the potential difference across the conductor to the current flowing through it (11 letters)
5. The SI unit of potential difference (4 letters)

Answers

3. Resistance.
4. Straight line.
5. Silver.
10. Resistivity is independent of physical dimensions of the conductor.
11. Aluminium.
12. Tungsten.
15. 5 A.
17. Manganin.
18. 2.5×10^{-8} Ω m.
19. 0.225 A.
20. 1/4.
21. 3.75 Ω.
22. 3 Ω.
23. 36 Ω.
26. 0.125 Ω.
27. $d = 2.67 \times 10^{-4}$ m.
28. (d)
29. (c)
30. (b)
31. (d)
32. (b)
33. (b)
34. potential difference, temperature
35. Resistance, ratio
36. friction
37. resistance
38. ohm meter
39. higher, 100
40. T.
41. F.
42. T.
43. F.
44. F.
45.
(1) [(a), (vi)]
(2) [(b), (viii)]
(3) [(c), (v)]
(4) [(d), (vii)]
(5) [(e), (i)]
(6) [(f), (iv)]
(7) [(g), (ii)]
(8) [(h), (iii)]
46.
(1) [(a), (iii)]
(2) [(b), (iv)]
(3) [(c), (vi)]
(4) [(d), (v)]
(5) [(e), (ii)]
(6) [(f), (i)]
47.
Across:
1. resistivity
2. series
3. parallel
6. ohm
Down:
4. resistance
5. volt

Hints/Solutions

15. Potential difference V = 12.5 V, resistance R = 2.5 Ω. Ohm's law gives
Current $I = V/R$ = 1.25 V / 2.5 Ω = 5A.

17. $R = \rho l / A$ gives $A = \rho l / R$, which shows that for same l and R and l, A is directly proportional to ρ. Now, manganin has higher value of ρ.

18. Radius r = 0.2 mm = 2×10^{-4} m, length l = 1.5 m, resistance R = 0.3 Ω.

Now, $R = \rho l / A$, which gives

$\rho = R A / l = R (\pi r^2) / l$

= $(0.3 \,\Omega) (3.14) (2 \times 10^{-4} \text{ m})^2 / (1.5 \text{ m}) = 2.5 \times 10^{-8}$ Ω m.

19. Here, potential differences are V_1 = 12 V, V_2 = 4.5 V.

Current I_1 = 0.6 A.

Let R be the resistance of conductor.

Ohm's law gives $V_1 = R \times I_1$, $V_2 = R \times I_2$.

Or $V_2/V_1 = I_2/I_1$

or $I_2 = (V_2/V_1) \times I_1 = (4.5 \text{ V}/12 \text{ V}) \times 0.6 \text{ A} = 0.225 \text{ A}$.

20. Since volume = mass /density of copper both A and B have the same mass and density. So, their volume is same. Or, $l_1 A_1 = l_2 A_2$, or, $(A_1/A_2) = (l_2/l_1) = 1/2$ (given).
$R_2/R_1 = (\rho l_2/A_2)/(\rho l_1/A_1) = (l_2/l_1)(A_1/A_2) = (1/2) \times (1/2) = 1/4$.

21. As in Solved Example 5.4, we get $R_2/R_1 = 1/4$, which gives
$R_2 = R_1/4 = 15.0 \ \Omega/4 = 3.75 \ \Omega$.

22. Let R_1, d, l and R_2, d_2, l_2 be the resistances, diameters and lengths of the first and the second wire respectively.
We have $R_1 = 6 \ \Omega$, $l_2 = 2 \ l$, $d_2 = 2 \ d$.
$R_1 = \rho l/A = \rho l/(\pi d^2/4)$,
$R_2 = \rho l_2/A_2 = \rho (2l)/\{\pi (2d)^2/4\}$
$= (1/2) \times (\rho l/(\pi d^2/4)) = R_1/2 = 6 \ \Omega/2 = 3 \ \Omega$.

23. Both stretched wire and original wire have the same volume.
So, $l_1 A_1 = l_2 A_2$, or, $(A_1/A_2) = (l_2/l_1) = 3$.
Now, $R_2/R_1 = (\rho l_2/A_2)/(\rho l_1/A_1) = (l_2/l_1)(A_1/A_2) = (3) \times (3) = 9$.
Or, $R_2 = 9 \times R_1 = 9 \times (4 \ \Omega) = 36 \ \Omega$.

26. Let R_1, d_1 and R_2, d_2 be the resistances and diameters of the first and the second wire respectively.
We have $R_1 = 2 \ \Omega$, $d_2 = 2 \ d_1$.
Ratio of areas of cross-section is
$(A_1/A_2) = (\pi d_1^2/4)/(\pi d_2^2/4) = (d_1/d_2)^2 = (1/2)^2 = 1/4$.
Both the wires have the same volume.
So, $l_1 A_1 = l_2 A_2$, or, $(l_2/l_1) = (A_1/A_2) = 1/4$.
Ratio of resistances is
$R_2/R_1 = (\rho l_2/A_2)/(\rho l_1/A_1) = (l_2/l_1)(A_1/A_2) = (1/4) \times (1/4) = 1/16$, which gives
$R_2 = (1/16) R_1 = (1/16) \times (2 \ \Omega) = 0.125 \ \Omega$.

27. Here, resistance $R = 7 \ \Omega$, length $l = 14$ m, resistivity $\rho = 2.8 \times 10^{-8} \ \Omega$ m
Diameter $d = ?$
$R = \rho l/A$ gives $A = \rho l/R$,
or $\pi d^2/4 = \rho l/R$,
or $d^2 = 4 \rho l/(\pi R)$
$= 4 \times (2.8 \times 10^{-8} \ \Omega \text{ m}) \times (14 \text{ m})/(3.14 \times 7 \ \Omega) = 156.8/21.98 = 7.13 \times 10^{-8}$ m^2, which gives
$d = (7.13 \times 10^{-8})^{1/2} = 2.67 \times 10^{-4}$ m.

28. Resistivity depends on the material of the conductor. Its value is independent of physical dimensions of the conductor.

29. The resistivity of an alloy like constantan being higher than that of metals, a small length of wire made of an alloy is sufficient to have required value of resistance.

30. $R = \rho l/A = \rho (2l)/(2A) = R'$.

31. Ohm's law is valid only if the temperature remains constant.

32. Current $I = V/R = 10 \text{ V} / 4 \text{ }\Omega = 2.5\text{A}$.

33. Resistivity is independent of physical dimensions of the conductor.

SUMMARY OF THE CHAPTER

- The resistance R of a conductor is directly proportional to its length l and inversely proportional to its area of cross-section A.
- R is proportional to (l/A), or $R = \rho\, l/A$, where ρ (Greek letter 'rho') is a constant of proportionality and is called the resistivity of the material of the conductor.
- Resistivity is an important property of the material of the conductor.
- The SI unit of resistivity is ohm meter (or Ω m).
- The resistivity of alloys is higher than that of metals by a factor of about 100.
- Copper and aluminium, being very good conductors, are used for electrical transmission lines.
- Wires made of alloys are generally used for making the heating elements of electric heating devices like electric iron, toaster and heater etc.

CHAPTER 6

RESISTANCE OF A COMBINATION OF RESISTORS

Combination of Resistances

Suppose we have a number of resistances, which, however, do not have the values we need for a particular application. In such a situation, we can combine the available resistances in different ways to obtain the desired values. We often come across such combinations in electric circuits used in modern electrical appliances.

There are two basic methods in which resistances can be combined. These are
(i) in series, and
(ii) in parallel.

Before studying them in some details, let us understand the important concept of equivalent resistance.

Equivalent resistance of a combination of Resistances

The equivalent resistance of a combination of resistances is the resistance of a single resistor, which can effectively replace the combination without producing any change in the current or potential difference in the rest of the circuit.

Resistors in series

Two or more resistances are said to be connected in series between two points when they are joined end to end so that there is only a single current path between these two points.

An example of the series combination of the resisters R_1, R_2 and R_3 is shown in Figure 6.1 in which the three resistors are connected end to end between the two points A and B. We notice that there exists a single path between A and B through which current can flow.

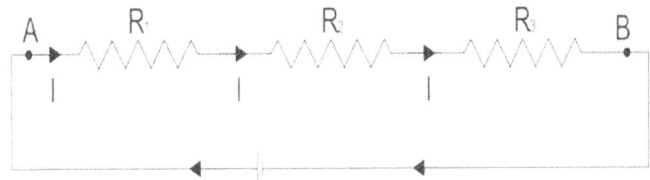

FIGURE 6.1 R_1, R_2 and R_3 in series

To learn more about the values of currents and potential differences in a series combination, we can carry out activity A 6.1

ACTIVITY A 6.1

The aim of this activity is study the current and potential difference relations for resistances in series.

The apparatus includes a battery of about 6 V (comprising of four dry cells of 1.5 V each), three different resistances R_1, R_2 and R_3 of a few ohms each, an ammeter, a voltmeter, a plug key and some thick copper wire for making connections.

The procedure consists of the following steps:

- Draw the circuit diagram as shown in Figure 6.2. Connect various components as per the circuit diagram with the help of thick copper wire. The three resistances R_1, R_2 and R_3 are joined in series. The positions of the ammeter and the voltmeter are as shown with solid lines in the diagram.

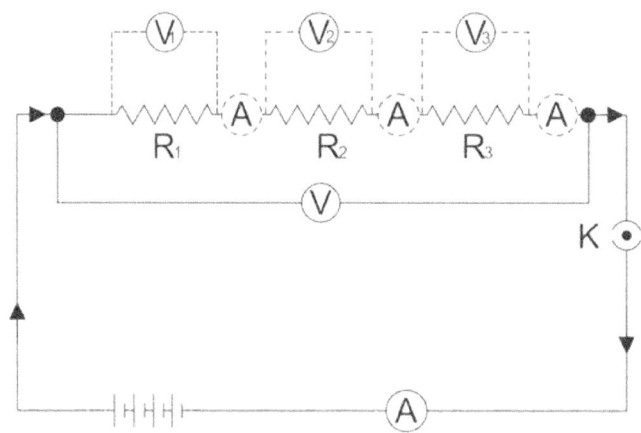

FIGURE 6.2 *The electric circuit to study the current and potential difference relations for resistances in series*

- Insert the key and note the readings of the ammeter and that of the voltmeter. Let these values be I and V respectively.
- Remove the key. Change the position of the ammeter and connect it between R_1 and R_2 as shown with dotted lines. Record the reading of the ammeter. Let it be I_1.
- Now, connect the ammeter between R_2 and R_3 and then between R_3 and the key. Each time, remove the key before changing the connection and reinsert it before taking the reading. Let the corresponding readings of the ammeter be I_2 and I_3 respectively.
- Next, connect the voltmeter across R_1 (as shown with dotted lines in Fig. 6.2) and note the corresponding reading (say V_1).
- Similarly, connect the voltmeter, first across R_2 and then across R_3. Let the corresponding readings of the voltmeter be V_2 and V_3 respectively.

Precaution: The key should be inserted into the plug only when taking the readings of voltmeter and ammeter and should be removed after these measurements.

What we observe in the above activity is as follows.

(i) The values of currents satisfy the following relation: $I = I_1 = I_2 = I_3$. It shows that the same current flows through every part of the circuit.
(ii) The values of potential differences are found to satisfy the following relation: $V = V_1 + V_2 + V_3$, which shows that in series combination of resistances, the total potential difference across the combination equals the sum of the potential differences across individual resistors.

Equivalent resistance of a series combination

To obtain the equivalent resistance of a series combination, we note that

(i) When resistors are connected in series, the same current flows through all of these resistors.
(ii) The total potential difference across the series combination equals the sum of the potential difference across individual resistors.

In Figure 6.3, V is the total potential difference between the ends A and B of the series combination. V_1, V_2 and V_3 are the potential difference across R_1, R_2 and R_3 respectively. We have

$$V = V_1 + V_2 + V_3 \quad (6.1)$$

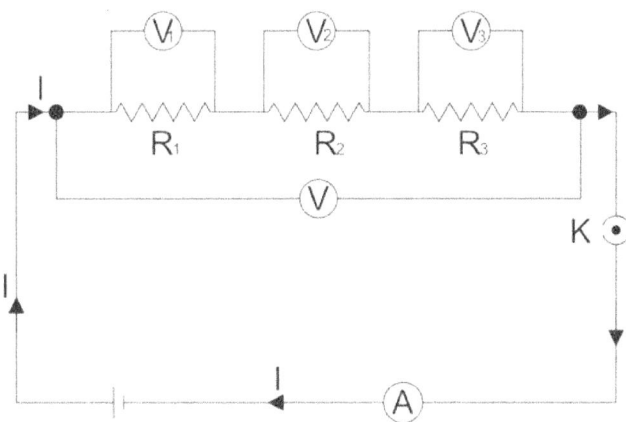

FIGURE 6.3 *Resistances in series*

Let *I* be the value of the common current in the circuit through each of the resistances. On applying Ohm's law to each of the resistors R_1, R_2 and R_3, we get

$$V_1 = I\,R_1 \quad (6.2(a))$$

$$V_2 = I\,R_2 \quad (6.2(b))$$

$$V_3 = I\,R_3 \quad (6.2(c))$$

Let R_s be the equivalent resistance of the combination. By definition, R_s is a single resistance which can replace the series combination in such a way that the values of potential difference *V* and the current *I* remain unchanged. (Fig. 6.4)

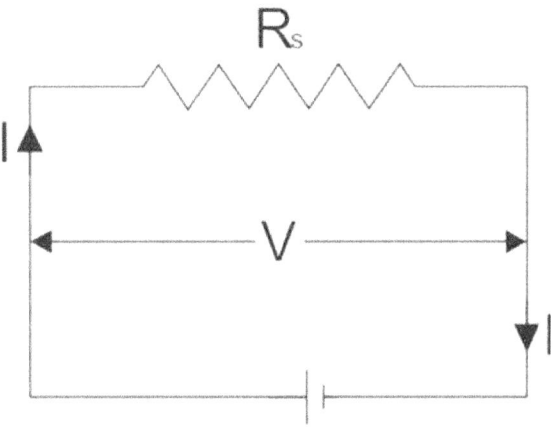

FIGURE 6.4

Again applying Ohm's law to the equivalent resistance, we have

$V = I R_s$ \hfill (6.3)

Substituting the values from Eqs. (6.2) and (6.3) in Eq. (6.1), we have

$I R_s = IR_1 + IR_2 + IR_3$

$R_s = R_1 + R_2 + R_3$ \hfill (6.4)

The above result leads to the following important conclusion. ***The equivalent resistance R_s of the resistances R_1, R_2 and R_3 in series equals the sum of the individual resistances.*** Although we have derived the above result for the case of three resistances, yet the nature of arguments used shows that same conclusion can be extended to any number of resistances in series.

Such a result tells us that ***the equivalent resistance of a series combination has a higher value than that of any individual resistances.***

SOLVED EXAMPLES

6.1 Resistors in series

Two resistors having resistance of 8 Ω and 2 Ω are joined in series with a source of constant potential difference. What will be the ratio of the potential differences across them?

Solution

Since the resistors are joined in series, same current I flows through both the resistors.

Therefore, $V_1 = R_1 I$, $V_2 = R_2 I$.

or, $V_2/V_1 = R_2/R_1 = (8\ \Omega / 2\ \Omega)$

So, $V_2/V_1 = 4$.

6.2 Resistors in series

An electric bulb having a filament of resistance 15 Ω is connected in series with a 6 Ω resistor. The combination is joined across a 12 V battery. Calculate

(i) the total resistance of the combination?

(ii) the current through the circuit and

(iii) the potential differences across the bulb and the resistor.

Solution

Resistance of the filament of the bulb, $R_1 = 15\ \Omega$

Resistance of the resistor, $R_2 = 6\ \Omega$.

(i) Since R_1 and R_2 are in series, the total resistance of the combination, R is

$R = R_1 + R_2 = 15\ \Omega + 6\ \Omega = 21\ \Omega$.

(ii) Potential difference across the series combination, $V = 12$ V.

Using Ohm's law, the current through the circuit

$I = V/R = 12\text{ V}/21\ \Omega = 0.57$ A.

(iii) Since they are joined in series, same current I flows through both the bulb and the resistor. Applying Ohm's law, separately to the bulb and the resistor, the potential difference across the bulb,

$V_1 = R_1 I = 15\ \Omega \times (12/21)\ A = 8.57\ V$.

The potential difference across the resistor is

$V_2 = R_2 I = 6\ \Omega \times (12/21)\ A = 3.43\ V$.

6.3 Resistors in series

Figure 6.5 shows two resistances of 6 Ω and 8 Ω along with an unknown resistance R joined in series across a battery of 24 V. The potential difference across the 6 Ω resistor is measured to be 6V. Calculate

(i) the current through the 6 Ω resistance,

(ii) the current through the unknown resistance,

(iii) the potential difference across the unknown resistance R,

(iv) the value of R, and

(v) the total resistance in the circuit.

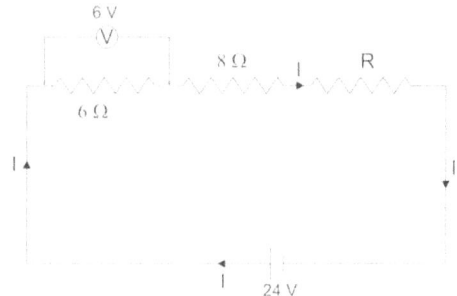

FIGURE 6.5

Solution

Let V_1, V_2 and V_3 be the voltage drops across $R_1 = 6\ \Omega$, $R_2 = 8\ \Omega$ and R respectively and V be the potential difference across the battery. We have

$V_1 = 6\ V,\ V_2 = ?\ V_3 = ?\ V = 24\ V$.

(i) Suppose I be the current through R_1. Applying Ohm's law to this resistance,

$V_1 = I R_1$

Or $I = V_1 / R_1 = 6\ V / 6\ \Omega = 1\ A$.

So, the current through R_1 is 1 A.

(ii) In a series circuit, a common current I flows through each of the resistors. So, the current through the unknown resistance and through R_2 is the same as through R_1 i.e., 1 A.

(iii) On applying Ohm's law to R_2, we get

$V_2 = R_2 I = 8\ \Omega \times (1A) = 8\ V$.

In a series circuit, total voltage across the combination equals the sum of voltage drops across the separate resistors.

So, $V = V_1 + V_2 + V_3$

Or, 24 V = 6 V + 8 V + V_3, which gives V_3 = 10 V.

Hence the potential difference across R is 10 V.

Using V_3 = 10 V, we get

R = 10 V/1 A = 10 Ω.

The value of the unknown resistance R is 10 Ω.

(iv) Let R_s be the total resistance in the series circuit. We have

$R_s = R_1 + R_2 + R$ = 6 Ω + 8 Ω + 10 Ω = 24 Ω.

Hence, the total resistance in the series circuit is 24 Ω.

6.4 Resistors in series

A battery of 12 V is connected in series with resistors of 1 Ω, 2 Ω, 3 Ω, 4 Ω and 5 Ω respectively. How much current would flow through the 4 Ω resistor?

Solution

Potential difference across the battery V = 12 V

In order to find the current through 4 Ω resistor, we need to find the value of current I through the series combination, which also equals the current through each of the resistances.

Equivalent resistances R_s in the series circuit is

$R_s = R_1 + R_2 + R_3 + R_4 + R_5$ = 1 Ω + 2 Ω + 3 Ω + 4 Ω + 5 Ω

or R_s = 15 Ω.

Applying Ohm's law to the whole circuit, we get

$V = R_s I$

or $I = V/R_s$ = 12 V/15 Ω = 0.8 A.

Hence, the current through 4 Ω resistance is 0.8 A.

6.5 Resistors in series

Four resistances R_1 = 2 Ω, R_2 = 3 Ω, R_3 = 4 Ω and R_4 = 7 Ω are joined in series across a battery of 12 V. What will be

(i) the current through 4 Ω resistance?

(ii) the potential difference across 3 Ω resistance?

Solution

(i) To find the current through 4 Ω resistor, we need to find the value of current I through the series combination, which also equals the current through each of the resistances.

Now total resistances R_s in the circuit is

$R_s = R_1 + R_2 + R_3 + R_4$ = 2 Ω + 3 Ω + 4 Ω + 7 Ω.

Or R_s = 16 Ω.

Applying Ohm's law to the whole circuit, we get

$V = R_s I$

Or $I = V/R_s$ = 12 V/16 Ω = 0.75 A.

Hence, the current through 7 Ω resistance is 0.75 A.

(ii) Let V_2 be the potential difference across R_2 (= 3 Ω) through which the current I flows.

Applying Ohm's law to R_2,

$V_2 = I R_2 = (0.75 \text{ A}) \times (3 \text{ Ω})$

or $V_2 = 2.25$ V.

EXERCISES E 6.1

Based on Resistors in Series

A. Very Short Answer Type Questions

1. Draw a circuit diagram showing a cell, a key and three resistances R_1, R_2 and R_3 joined in series.
2. Write the formula for the equivalent resistance of three resistances R_1, R_2 and R_3 joined in series.
3. Does a common potential difference exist across each of the resistances in series?
4. Does a common current flow through each of the resistors in series?

B. Short Answer Type Questions – I

5. Derive an expression for the common current I in the circuit in terms of the two resistors R_1 and R_2 joined in series with a cell of voltage V.
6. *(Numerical Problem Based on Higher Order Thinking Skills (HOTS))* Two resistors, whose resistances are in the ratio 1:4, are joined in series with a source of constant potential difference. What will be the ratio of the potential differences across them?

C. Short Answer Type Questions – II

7. What is the equivalent resistance of a combination of resistances? When are two resistors R_1 and R_2 said to be joined in series?
8. Draw a circuit diagram showing three resistors R_1, R_2 and R_3 joined in series with a cell and an ammeter. Also show a voltmeter which reads the potential difference across the resistance R_2 and another voltmeter which reads the potential difference across the combination.
9. *(Numerical Problem)* An electric bulb having resistance of 10 Ω and a resistor of resistance 2 Ω are connected in series with a 12 V battery. What will be
(a) the total resistance of the circuit?
(b) the current through the bulb?
(c) the potential difference across the bulb?
(d) the current through the resistor?
10. *(Numerical Problem)* Consider the circuit in Figure 6.6. Calculate

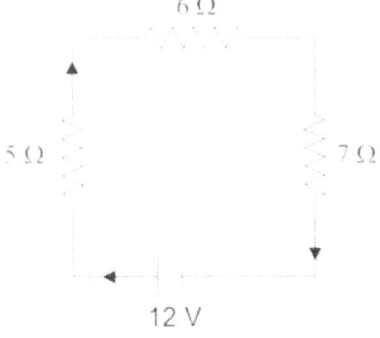

FIGURE 6.6

(a) the current through each of the resistors.

(b) the potential difference across each of resistors.

11. *(Numerical Problem)* In the circuit shown in Figure 6.7, the values of resistances are $R_1 = 5\ \Omega$, $R_2 = 7\ \Omega$ and $R_3 = 12\ \Omega$. The voltmeter V shows a reading of 3 V. Calculate
(a) the current through R_1, R_2 and R_3
(b) the potential difference across R_1, R_2 and R_3 and

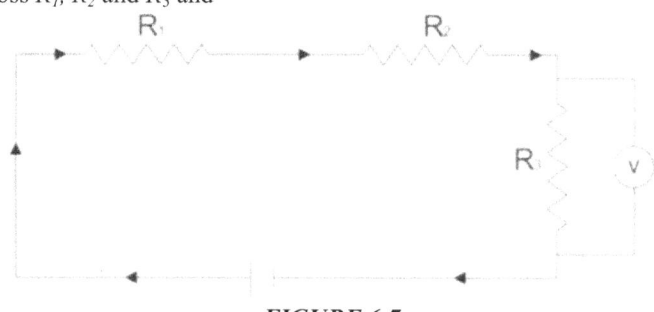

FIGURE 6.7

(c) the potential difference across the battery.

D. Long Answer Type Questions

12. With the help of a suitable diagram, derive the value of equivalent resistance of three resistors R_1, R_2 and R_3 joined in series?

E. Objective Questions

I. Multiple Choice Type Questions

Choose the Correct Answer:

13. To obtain maximum resistance by joining the given resistors, they should be grouped in:
(a) series
(b) parallel
(c) mixture of series and parallel combinations.
(d) none of the above.

14. Three resistances of 15 Ω, 5 Ω, 1 Ω, respectively are connected in series. New resistance will be:
(a) between 5 Ω and 2 Ω
(b) between 10 Ω and 5 Ω
(c) more than 15 Ω
(d) less than 2 Ω

15. Three unequal resistors are connected in series across a cell. Which of the following statement is true?
(a) Current through smallest resistor is maximum
(b) Same current flows in all the resistors
(c) Current through largest resistor is maximum
(d) Any one of the above can be true depending on the potential difference of the cell.

16. Two unequal resistors are connected in series with a cell. Which of the following statement is true?
(a) Potential drop across both resistors is same
(b) Potential drop across smaller resistor is more
(c) Potential drop across larger resistor is more
(d) Any one of the above can be true depending on the potential difference of the cell.

17. The equivalent resistance in series combination is
(a) smaller than the largest resistance
(b) larger than the smallest resistance
(c) smaller than the smallest resistance
(d) larger than the largest resistance.

II. Fill in the blanks Type

Fill in the blanks:

18. The .... resistance of a combination of resistances is a single resistance, which can effectively replace the combination without producing any change in the ...... or potential .... in the rest of the circuit.

19. The total potential difference across the .... combination equals the sum of the potential difference across separate resistors.

20. In case of ..... circuit, there is a common path through which the current flows.

III. True or False Type

Mark the following Statements True (T) or False (F)

21. The total current flowing through the series combination equals the sum of individual currents through different resistances. T/F

22. While using a number of devices in a series circuit, failure of one of them due to some defect will result in the breaking of the circuit. T/F

Answers

3. No.
4. Yes.
6. 1:4.
9. (a) 12 Ω (b) 1.0 A (c) 10.0 V (d) 1.0 A
10. (a) 0.67 A (b) 3.33 V, 4.0 V, 4.67 V.
11. (a) 0.25 A (b) 1.25 V, 1.75 V (c) 6.0 V.
13. (a)
14. (c)
15. (b)
16. (c)
17. (d)
18. equivalent, current, difference
19. series
20. series
21. F.
22. T.

Hints/Solutions

6. The two resistances can be taken as R and $4R$ respectively. Since the resistors are joined in series, same current I flows through both the resistors. $V_1 = RI$, $V_2 = (4R)I$. So, $V_1/V_2 = 1/4$.

9. $R_1 = 10\ \Omega$, $R_2 = 2\ \Omega$, $V = 12$ V. (a) The total resistance $R_s = R_1 + R_2 = 12\ \Omega$ (b) $I = V/R_s = 12\text{ V}/12\ \Omega = 1.0$ A (c) $V_1 = R_1 I = 10\ \Omega \times 1.0$ A $= 10.0$ V (d) same as in (b).

10. $R_1 = 5\ \Omega$, $R_2 = 6\ \Omega$, $R_3 = 7\ \Omega$, $V = 12$ V. The total resistance $R_s = R_1 + R_2 + R_3 = 18\ \Omega$. (a) in a series circuit, a common current I flows through each of the resistors. $I = V/R_s = 12\text{ V}/18\ \Omega = (2/3)$ A. (b) $V_1 = I R_1 = 5\ \Omega \times (2/3)$ A $= 3.33$ V, $V_2 = R_2 I = 6\ \Omega \times (2/3)$ A $= 4.0$ V, $V_3 = R_3 I = 7\ \Omega \times (2/3)$ A $= 4.67$ V.

Note: We find that $V = V_1 + V_2 + V_3$, which is expected for a series combination.

11. $R_1 = 5\ \Omega$, $R_2 = 7\ \Omega$, $R_3 = 12\ \Omega$, $V_3 = 3$ V. (a) current through R_3 is $I_3 = V_3/R_3 = 3\text{ V}/12\ \Omega = 0.25$ A. In a series circuit, a common current I flows through each of the resistors. So, current through R_1 and R_2 is also 0.25 A.

(b) $V_1 = R_1 I = 5\ \Omega \times 0.25$ A $= 1.25$ V, $V_2 = R_2 I = 7\ \Omega \times 0.25$ A $= 1.75$ V.

(c) In a series combination, the potential difference across the combination equals the sum of the potential difference across separate resistors. So, potential difference across the battery is $V = V_1 + V_2 + V_3 = 1.25$ V $+ 1.75$ V $+ 3$ V $= 6.0$ V.

13. The equivalent resistance of a series combination is obtained simply by adding the values of the individual resistances. Obviously, such a sum gives the maximum value.

14. The equivalent resistance of a series combination is obtained simply by adding the values of the individual resistances. Obviously, such a sum gives a value, which is larger than any individual value.

15. When resistors are connected in series, same current flows through every part of the circuit.

16. In series, the same current flows through each resistor. So, Ohm's law tells that potential drop is proportional to the value of the resistance.)

17. The equivalent resistance of a series combination being the sum of resistances is larger than any individual value.

Resistors in Parallel

Two or more resistances are said to be connected in parallel between two points A and B when one end of each of them is joined to A and the other end of each of them is joined to B so that a common potential difference (= $V_A - V_B$) exists across each one of them.

Figure 6.8 shows a parallel combination of the resisters R_1, R_2 and R_3. One end of all the three resistors is connected to point A, which is further joined to the positive terminal and the other end of all of them is connected to B, which is further joined to the negative terminal of the battery. We notice that the potential difference across each of them equals the potential difference between the two points A and B.

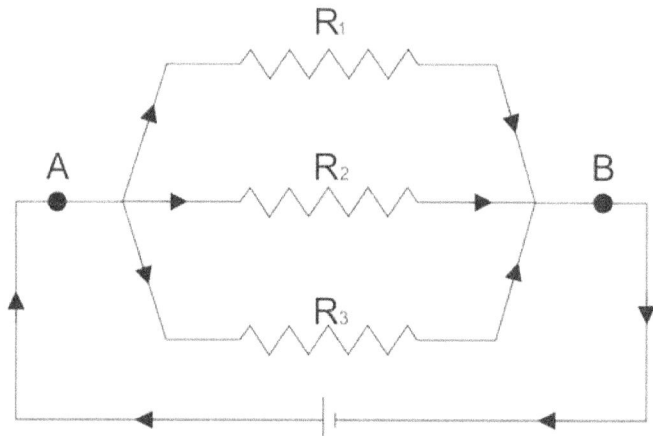

FIGURE 6.8 R_1 R_2 and R_3 in parallel

Let us carry out the following activity to learn more about the values of currents and potential differences in a parallel combination.

ACTIVITY A 6.2

The aim of this activity is study the current and potential difference relations for resistances in a parallel combination.

The apparatus required includes a battery of about 6 V (comprising of four dry cells of 1.5 V each), three resistances R_1, R_2 and R_3 of different values of around a few ohms each, a voltmeter, an ammeter, a plug key and some thick copper wire for making connections.

The procedure consists of the following steps:

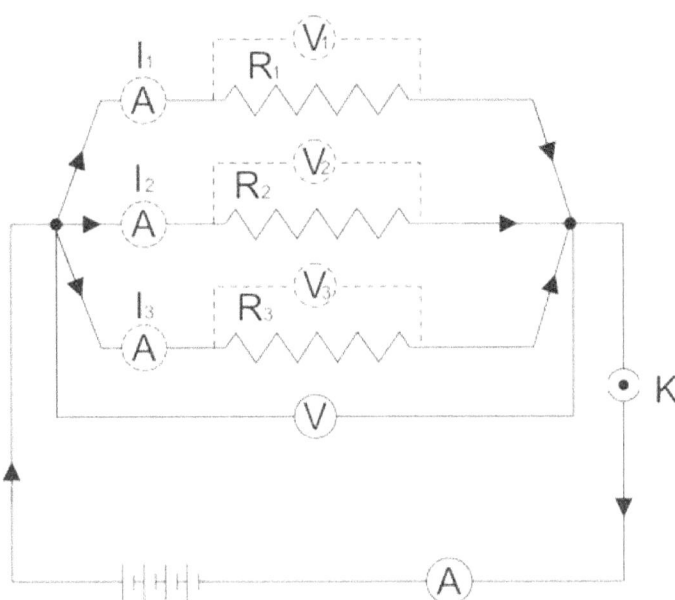

FIGURE 6.9 *The electric circuit to study the current and potential difference relations for resistances in parallel*

- Draw the circuit diagram as shown in Figure 6.9. Connect various components as per the circuit diagram with the help of thick copper wire. The three resistances R_1, R_2 and R_3 are joined in parallel. The positions of voltmeter and the ammeter are such that they read the values of potential difference across and the total current through the parallel combination respectively.
- Insert the key and note the readings of the ammeter and that of the voltmeter. Let these values be I and V respectively.
- Remove the key. Change the position of the voltmeter and connect it across R_1 as shown with dotted line in Figure 6.9. Record the reading (say V_1) of the voltmeter.
- Now, connect the voltmeter across R_2 and then across R_3. In each case, the key should be removed before changing the connection and reinserted before taking the reading. Let the corresponding readings of the voltmeter be V_2 and V_3 respectively.
- Next, connect the ammeter in series with R_1 (as shown dotted in Figure 6.9) and note the corresponding reading (say I_1).
- Similarly, connect the ammeter, first in series with R_2 and then in series with R_3. Let the corresponding readings of the ammeter be I_2 and I_3 respectively.

Precaution: Same as in Activity A 6.1.

What we observe in the above activity is as follows.

(i) The values of potential differences satisfy the relation $V = V_1 = V_2 = V_3$, which shows that in a parallel combination, a common potential difference exists across each of the resistances.
(ii) The values of currents are found to satisfy the following relation: $I = I_1 + I_2 + I_3$. This relation result shows that in a parallel combination of resistances, the total current flowing through the combination equals the sum of individual currents through different resistances.

Equivalent resistance of a parallel combination

To obtain the equivalent resistance of a parallel combination of resistances, we note that

(i) A common potential difference exists across each of the resistances joined in a parallel combination.
(ii) The total current flowing through the parallel combination equals the sum of individual currents through different resistances.

In Figure 6.10, I is the total current through the parallel combination and I_1, I_2 and I_3 are the currents through R_1, R_2 and R_3 respectively.

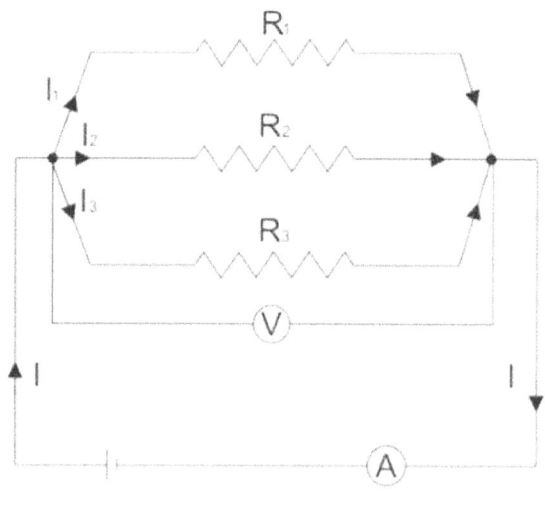

FIGURE 6.10

We have

$$I = I_1 + I_2 + I_3 \qquad (6.5)$$

Let V be the common potential difference across each of the resistances. On applying Ohm's law to the resistors R_1, R_2 and R_3, we get

$$I_1 = V/R_1 \qquad (6.6\ (a))$$

$$I_2 = V/R_2 \qquad (6.6\ (b))$$

$$I_3 = V/R_3 \qquad (6.6\ (c))$$

Let R_p be the equivalent resistance of the combination. By definition, R_p is a single resistance which can replace the parallel combination in such a way that the values of potential difference V and the current I remain unchanged.

Again by applying Ohm's law to the equivalent resistance (Figure 6.11), we have

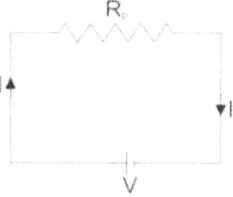

FIGURE 6.11

$$I = V/R_p \qquad (6.7)$$

From Eqs. (6.5), (6.6) and (6.7), we have

$V/R_p = V/R_1 + V/R_2 + V/R_3$

or $1/R_p = 1/R_1 + 1/R_2 + 1/R_3$

(6.8)

The above results lead to the following important conclusion: *the reciprocal of equivalent resistance R_p of a parallel combination equals the sum of the reciprocals of individual resistances R_1, R_2 and R_3.*

Although we have derived the above result for the case of three resistances in parallel, yet the same conclusion holds true for any number of resistances in parallel.

Such result tells us that *the equivalent resistance of a parallel combination has a lower value than that of any individual resistances.*

When there are only two resistors, use of Eq. (6.8) gives

$1/R_p = 1/R_1 + 1/R_2$

$= (R_1 + R_2)/R_1 R_2$

or $R_p = R_1 R_2 / (R_1 + R_2)$

(6.9)

Eq. (6.8) gives us the direct result for the case of two resistances.

Advantages of a parallel circuit over a series circuit

For the purpose of using electric devices, a parallel circuit has many advantages over the series circuit. We list below some important ones.

(i) Simultaneous use of different types of electric devices

Electric devices used for different applications need different values of current to operate properly. For example, heating devices like electric heater and electric geyser require considerably higher value of current than, say, an ordinary electric bulb. Such heating devices and the bulb cannot be used simultaneously in a series circuit, because a common current flows through the whole circuit. On the other hand, in case of a parallel circuit, the current values are different in different branches, even though a common potential difference exists across each branch. Devices having different current ratings but same voltage rating can, therefore, be connected in different branches of parallel circuit.

(ii) Uninterrupted working of devices in good working condition

In case of series circuit, there is a common path through which the current flows. So, while using a number of devices in a series circuit, failure of one of them due to some defect will result in the breaking of the circuit. Consequently none of the other devices will work even though they may be in good working condition. Such a situation sometimes arises on festival occasions, where tiny bulbs are used in series for decoration purpose. If one of them fails, all the bulbs stop working. The bulbs would start working again only when one locates the defaulting bulb and replaces it with a good one.

Such a problem does not arise in case of a parallel circuit. In a parallel combination, the current flows simultaneously in different branches. So stopping of current in one branch due to a defective device does not result in stopping of current in other branches. As a result, the other devices in good condition keep on working.

(iii) Higher Value of total current through the circuit

We know the equivalent resistance of parallel combination is low. As a result we can obtain higher value of total current I from the power supply line of a given voltage (recall that I is proportional to $1/R$ for constant V). Also, in the parallel circuit, the total current gets divided into various branches. So, the higher value of total current makes it possible for various electric devices in different branches to have sufficient value of current for their proper functioning. On the other hand, the equivalent resistance of a series circuit being high, the total current from a given power supply has a lower value.

The same low current passes through all the electric devices in the series circuit. Many a time, the low value of current in is too small for the proper working of some of these devices.

SOLVED EXAMPLES

6.6 Resistors in Parallel

(Based on Higher Order Thinking Skills (HOTS))

How many 44 Ω resistors need to be joined in parallel to total resistance of 5.5 Ω?

Solution

Let the number of resistors required = n

Resistance of each resistor $R = 44\ \Omega$

Let R_p be the equivalent resistance of this parallel combination. Then

$1/R_p = 1/R + 1/R + \ldots n\ times = n/R$

or $R_p = R/n = 44\ \Omega/n$

So, $5.5\ \Omega = 44\ \Omega/n$

or $n = 44\ \Omega/5.5 = 8$

Hence, required number of resistors = 8.

6.7 Resistors in Parallel

In Figure 6.12 (a) given below, five resistances $R_1 = 16\ \Omega$, $R_2 = 80\ \Omega$, $R_3 = 8\ \Omega$, $R_4 = 36\ \Omega$ and $R_5 = 24\ \Omega$ together with a battery of 18 V are connected as shown. Calculate

(i) the total resistance in the circuit

(ii) the total current flowing in the circuit?

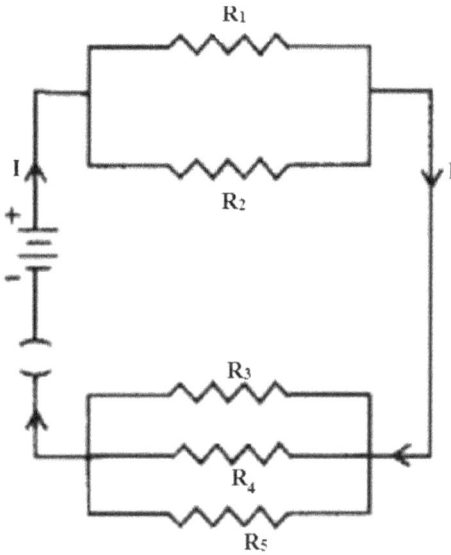

FIGURE 6.12 (a)

Solution

(i) The two resistances R_1 and R_2 are in parallel. Let R' be the equivalent resistance of this combination. Then,

$1/R' = 1/R_1 + 1/R_2 = 1/16 + 1/80 = 3/40$

or $R' = (40/3)\ \Omega$.

Again, three resistances R_3, R_4 and R_5 are in parallel. Let R'' be the equivalent resistance of this parallel combination. Then

$1/R'' = 1/R_3 + 1/R_4 + 1/R_5 = 1/8 + 1/36 + 1/24 = 14/72$

Or $R'' = (36/7)\ \Omega$.

The circuit can be redrawn as shown in Figure 6.12 (b), where R' and R'' are seen to be joined in series. The total resistance R_T is given by

$R_T = R' + R'' = (40/3)\ \Omega + (36/7)\ \Omega = (388/21)\ \Omega$.

i.e., $R_T = 18.48\ \Omega$.

So, the total resistance in the circuit is $18.48\ \Omega$.

FIGURE 6.12 (b)

(ii) To calculate the total current in the circuit, we can apply Ohm's law to the equivalent circuit (Figure 6.12 (c)) to get

$I = V/R_T = (18\ \text{V}) / (18.48\ \Omega)$.

Or $I = 0.97\ \text{A}$.

FIGURE 6.12 (c)

So the total current in the circuit is 0.97 A.

6.8 Resistors in Parallel

A resistor having a resistance of 5 Ω is joined in parallel with an unknown resistance R. The effective resistance of the parallel combination is found to be 3 Ω. What will be the value of equivalent resistance if the same resistors are joined in a series combination?

Solution

R_1 (= 5 Ω) and R are in parallel.

The equivalent resistance of the parallel combination R_p = 3 Ω.

We have, $1/R_p = 1/R_1 + 1/R$, which gives

$1/R = 1/R_p - 1/R_1 = 1/3 - 1/5 = 2/15$.

Or $R = (15/2)$ Ω = 7.5 Ω.

The equivalent resistance of the series combination is

$R_s = R_1 + R = 5$ Ω + 7.5 Ω = 12.5 Ω.

6.9 Resistors in Parallel

An electric bulb of resistance 50 Ω, an electric fan of resistance 100 Ω and an electric iron of resistance 500 Ω are connected in parallel to a 200 V source. Calculate

(i) the currents drawn by each of these appliances?

(ii) the total resistance in the circuit

(iii) the total current flowing in the circuit?

Solution

We have $R_1 = 50$ Ω, $R_2 = 100$ Ω and $R_3 = 500$ Ω.

(i) The common potential difference across each of the resistances in parallel is V = 200 V. On applying Ohm's law to the resistors R_1, R_2 and R_3, we get

$I_1 = V/R_1 = (200 \text{ V}) / (50 \text{ Ω}) = 4$ A

$I_2 = V/R_2 = (200 \text{ V}) / (100 \text{ Ω}) = 2$ A

$I_3 = V/R_3 = (200 \text{ V}) / (500 \text{ Ω}) = 0.4$ A.

(ii) Let R_p be the equivalent resistance of the parallel combination. Then

$1/R_p = 1/R_1 + 1/R_2 + 1/R_3$

$= 1/50 + 1/100 + 1/500 = 16/500 = 4/125$.

Or $R_p = (125/4)$ Ω = 31.25 Ω.

(iii) Let I_T be the total current drawn by the combination. Applying Ohm's law to the combination,

$I_T = V/R_p = (200 \text{ V}) / (31.25 \text{ Ω}) = 6.4$ A.

(**Note:** We find that $I_T = I_1 + I_2 + I_3$, as expected.)

6.10 Resistances in Series and Parallel Combinations

(Based on Higher Order Thinking Skills (HOTS))

How can three resistors of resistance 2 Ω, 3 Ω and 6 Ω be connected to give a total resistance of (a) 11 Ω (b) 1 Ω?

Solution

(a) First, note that in series combination the equivalent resistance is larger the largest resistance. So, to get a resistance of 11 Ω, we need to join the three resistances in series.

Equivalent resistances R_s in the series circuit is

$R_s = R_1 + R_2 + R_3 = 2$ Ω + 3 Ω + 6 Ω = 11 Ω.

(b) To get a resistance of 1 Ω, note that in a parallel combination, the equivalent resistance is smaller than the smallest resistance of 2 Ω. So, we need to join the three resistances in parallel. If R_p be the equivalent resistance of the parallel combination, we have

$1/R_p = 1/2 + 1/3 + 1/6 = (6/6)$, which gives $R_p = 1$ Ω, as desired.

6.11 Resistances in Series and Parallel Combinations
(Based on Higher Order Thinking Skills (HOTS))

What is (a) the highest (b) the lowest total resistance that can be obtained by combining three coils of resistance values $R_1 = 2$ Ω, $R_2 = 3$ Ω and $R_3 = 7$ Ω?

Solution

(a) The equivalent resistance of a series combination is obtained simply by adding the values of the individual resistances. Obviously, such a sum gives the highest value. So, to obtain the highest value, we need to join the resistances in series. Let R_s be the equivalent resistance of the series combination. We have

$R_s = R_1 + R_2 + R_3 = 2$ Ω $+ 3$ Ω $+ 7$ Ω $= 12$ Ω.

Hence, the highest value that can be obtained by combining the coils is 12 Ω.

(b) We know that the value of equivalent resistance of a parallel combination is smaller than the smallest of the individual resistances. Obviously, such a result gives the lowest value among all possible combinations.

So, to obtain the lowest value, we need to join all the resistances in parallel.

Let R_p be the equivalent resistance, we have

$1/R_p = 1/R_1 + 1/R_2 + 1/R_3$

$= 1/2 + 1/3 + 1/7 = 41/42$

or $R_p = (42/41)$ Ω $= 1.02$ Ω.

Hence, the lowest value that can be obtained by combining the coils is 1.02 Ω.

6.12 Resistances in Series and Parallel Combinations
(Based on Higher Order Thinking Skills (HOTS))

Six one ohm resistances are connected to form a regular hexagon as shown in Figure 6.13(a). Calculate the resistance of the combination if the current enters at the point A and leaves at the point B.

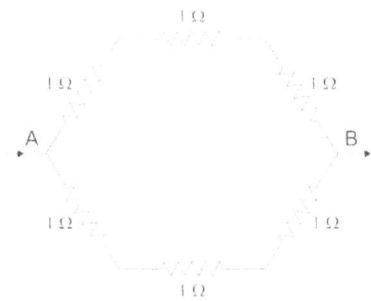

FIGURE 6.13 (a)

Solution

In the given electric circuit, a series combination of three resistances of 1 Ω each is joined in parallel to another series combination of three resistance of 1Ω each. Let R' and R'' be the equivalent resistance of the two series combinations. We have,

$R' = 1\,\Omega + 1\,\Omega + 1\,\Omega = 3\,\Omega$.

Obviously, $R'' = R' = 3\,\Omega$.

The circuit can be redrawn as in Figure 6.13 (b). Now, R' and R'' are joined in parallel.

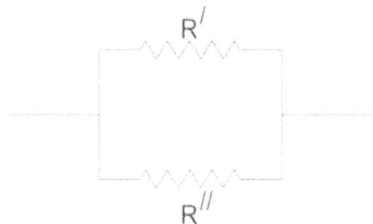

FIGURE 6.13 (b)

Let R_T be the total resistance of the combination. Then,

$R_T = R'\, R'' / (R' + R'') = (3\,\Omega) \times (3\,\Omega) / (3\,\Omega + 3\,\Omega) = 1.5\,\Omega$.

6.13 Resistances in Series and Parallel Combinations

(Based on Higher Order Thinking Skills (HOTS))

When two resistors are joined in series, the equivalent resistance is 18 Ω. When the same resistors are joined in parallel, the equivalent resistance is 4 Ω. Find the resistances of individual resistors.

Solution

We have, $R_s = R_1 + R_2 = 18\,\Omega$, which gives

$R_2 = 18 - R_1$

Again, $R_p = R_1 \times R_2 / (R_1 + R_2) = 4\,\Omega$, which gives

and $R_1 \times R_2 / (18) = 4$,

or $R_1 \times (18 - R_1) = 72$,

or $18\,R_1 - R_1^2 = 72$, which gives the equation,

$R_1^2 - 18\,R_1 + 72 = 0$,

or $(R_1 - 12)(R_1 - 6) = 0$.

Therefore, either, $R_1 = 12$, which gives $R_2 = 18 - R_1 = 6$,

or $R_1 = 6$, which gives $R_2 = 18 - R_1 = 12$.

Hence, the two resistances have values of 12 Ω and 6 Ω.

6.14 Resistances in Series and Parallel Combinations

An electric heater has two resistance coils of 15 Ω each, which may be used either in series or in parallel. What are the currents in the two cases when the heater is connected to a 220 V line?

Solution

Line voltage $V = 220$ V

When the coils are used in series,

Equivalent resistance of the series combination

$R_s = R_1 + R_2 = 15\ \Omega + 15\ \Omega = 30\ \Omega$

Using Ohm's law, current through the series combination

$I_s = V/R_s = 220\ \text{V}/30\ \Omega = 7.33$ A.

When the coils are used in parallel,

Equivalent resistance of the parallel combination R_p is given by

$R_p = R_1 \times R_2 / (R_1 + R_2) = (15\ \Omega \times 15\ \Omega)/(15\ \Omega + 15\ \Omega) = 7.5\ \Omega$

Again, using Ohm's law, current through the parallel combination

$I_p = V/R_p = 220\ \text{V}/(7.5\ \Omega) = 29.33$ A.

6.15 Resistances in Series and Parallel Combinations
(Based on Higher Order Thinking Skills (HOTS))

Five one ohm resistances are connected in the form a letter "A" as shown in Figure 6.14 (a). Calculate the equivalent resistance between the ends X and Y.

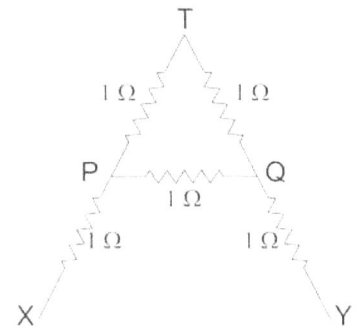

FIGURE 6.14 (a)

Solution

The circuit can be redrawn as in Figure 6.14 (b).

FIGURE 6.14 (b)

Between the points P and Q, we have two resistances of 1 Ω each in series joined in parallel to another resistance of 1 Ω. The equivalent resistance of the two resistances in series is

= 1 Ω + 1 Ω = 2 Ω. So, the equivalent resistance between P and Q is

$R' = (2\,\Omega) \times (1\,\Omega) / (2\,\Omega + 1\,\Omega) = (2/3)\,\Omega$. Hence, the equivalent resistance of the whole combination between the ends X and Y (Figure 6.14 (c)) is

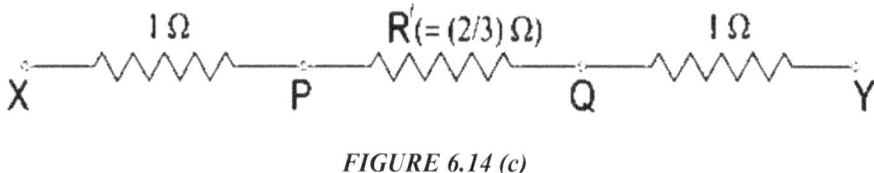

FIGURE 6.14 (c)

$R = 1\,\Omega + (2/3)\,\Omega + 1\,\Omega = (8/3)\,\Omega = 2.67\,\Omega$.

EXERCISES E 6.2

Based on Resistors in Parallel and Series and Parallel Combination

A. Very Short Answer Type Questions

1. Draw a diagram showing these resistances R_1, R_2 and R_3 joined in parallel.
2. Write the formula for the equivalent resistance of three resistances R_1, R_2 and R_3 joined in parallel with each other.
3. Does a common current flow through each of the resistors in parallel?
4. Does a common potential difference exist across each of the resistances in parallel?
5. Is the equivalent resistance R_p of two resistances R_1 and R_2 joined in parallel less than or more than the individual resistances R_1 and R_2?

B. Short Answer Type Questions – I

6. What is the meaning of a parallel combination of two or more resistances? How is the potential difference across the parallel combination related to the potential differences across individual resistors?
7. When are two resistances R_1 and R_2 said to be joined in parallel? Draw a diagram showing the parallel combination of R_1 and R_2. How is the total current through the parallel combination related to the currents through individual resistors R_1 and R_2?
8. Draw a circuit diagram showing a combination of three resistances R_1, R_2 and R_3 joined in parallel with each other together with a battery of voltage V joined across the combination and an ammeter which measures the current through R_2. Also show an ammeter which reads the current through the combination.
9. Which type of combination of resistances should be used to obtain an equivalent resistance lower in value than the two individual resistors R_1 and R_2? Write the formula for the equivalent resistance of the desired combination.
10. List two reasons why it is advantageous to connect electrical gadgets in parallel with the electric supply?

C. Short Answer Type Questions – II

11. *(Numerical Problem)* Three resistances R_1, R_2 and R_3 having values 6 Ω, 12 Ω and 36 Ω are connected in parallel across a battery of 24 V. Calculate
(a) the total resistance
(b) the total current in the circuit and
(c) the current through each resistance.
12. *(Numerical Problem)* A resistor having a resistance of 5 Ω is joined in series with an unknown resistance R. The effective resistance of the series combination is found to be 20 Ω. What will be the value of equivalent resistance if the same resistors are joined in a parallel combination?

D. Long Answer Type Questions

13. Obtain the equivalent resistance R_p of three resistances R_1, R_2 and R_3 joined in parallel with a battery?

14. *(Numerical Problem Based on Higher Order Thinking Skills (HOTS))* In the circuit diagram given below (Figure 6.15), a combination of resistors are joined across a battery of 6 V. Find
(a) the equivalent resistance of combination
(b) the total current drawn by the battery
(c) the currents through each of the four resistors
(d) the potential difference across each resistor.

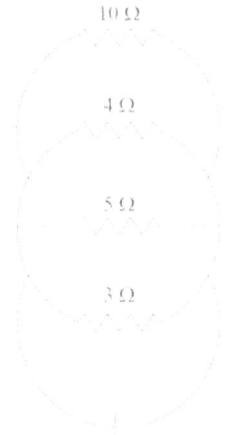

FIGURE 6.15

15. *(Numerical Problem Based on Higher Order Thinking Skills (HOTS))* Calculate the equivalent resistance between the points A and B in figure 6.16.

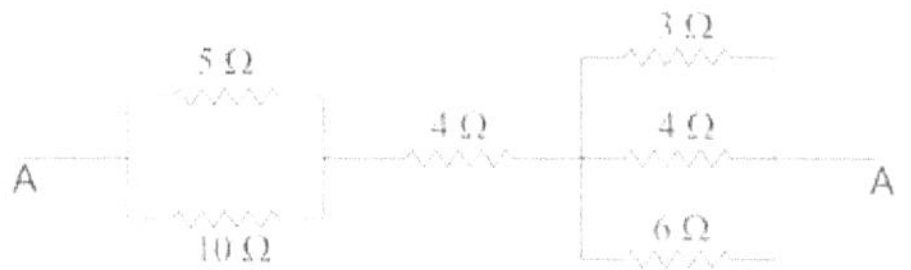

FIGURE 6.16

16. *(Numerical Problem Based on Higher Order Thinking Skills (HOTS))* When two resistors R_1 and R_2 are joined in series, the equivalent resistance is 25 Ω. When the same resistors are joined in parallel, the equivalent resistance is 6 Ω. Find R_1 and R_2.

17. *(Numerical Problem Based on Higher Order Thinking Skills (HOTS))* In the circuit shown below (Figure 6.17), the ammeter reads 1.25 A. Find

(a) the current through each of the resistors having resistances of 5 Ω and 10 Ω respectively

(b) the current through 20 Ω resistance

(c) the reading of the voltmeter V.

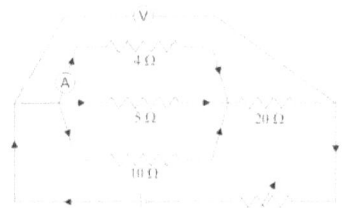

FIGURE 6.17

18. *(Numerical Problem Based on Higher Order Thinking Skills (HOTS))* The Figure 6.18 given below shows a part of a circuit. The voltmeter shows a reading of 12 V. Find (a) the current through each of the 5 Ω and 10 Ω resistors (b) reading of the ammeter A.

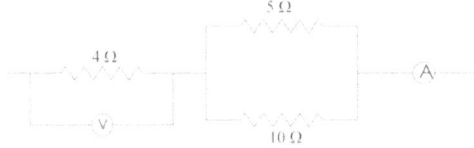

FIGURE 6.18

E. Objective Questions

I. Multiple Choice Type Questions

Choose the Correct Answer:

19. Three resistances of 10 Ω, 5 Ω and 2 Ω respectively are connected in parallel. The equivalent resistance will be
(a) between 10 Ω and 5 Ω
(b) more than 10 Ω
(c) less than 2 Ω
(d) between 5 Ω and 2 Ω

20. To obtain minimum resistance by joining the given resistors, they should be grouped in:
(a) parallel
(b) series
(c) mixture of series and parallel combinations.
(d) none of the above

21. Three resistances of 12 Ω, 6 Ω, 1 Ω, respectively are connected in parallel. the equivalent resistance will be:
(a) between 6 Ω and 1 Ω
(b) between 12 Ω and 6 Ω
(c) more than 12 Ω
(d) less than 1 Ω

22. *(Based on Higher Order Thinking Skills (HOTS))* Three unequal resistors are connected in parallel across a cell. Which of the following statement is true?
(a) Current through smallest resistor is maximum
(b) Same current flows in all the resistors
(c) Current through largest resistor is maximum
(d) Any one of the above can be true depending on the potential difference of the cell.

23. The equivalent resistance in parallel combination is
(a) smaller than the largest resistance
(b) larger than the smallest resistance
(c) smaller than the smallest resistance
(d) larger than the largest resistance.

24. *(Based on Higher Order Thinking Skills (HOTS))* The smallest resistance that can be obtained by the combination of n resistors, each of resistance R is:

(a) R/n^2
(b) R/n
(c) nR
(d) n^2R.

II. Fill in the blanks Type

Fill in the blanks:

25. The …….. of equivalent resistance of a …… combination equals the sum of the reciprocals of individual resistances.

26. Devices having different current ratings can be connected in different branches of a …….. circuit.

III. True or False Type

Mark the following Statements True (T) or False (F)

27. While using a number of devices in a parallel circuit, failure of one of them due to some defect will result in the breaking of the circuit. T/F

28. In a parallel combination, we can obtain higher value of total current from the power supply line of a given voltage. T/F

IV. Matching Type Questions

29. Match these items in column A to the corresponding terms in column B. Note that more than one item in column A may match with the same item in column B.

S. No.	Column A	Column B
1.	(a) A single resistance, which can effectively replace the combination of resistances.	(i) a parallel combination of resistances
2.	(b) The equivalent resistance equals the sum of the individual resistances.	(ii) parallel circuit
3.	(c) total current through the combination equals the sum of individual currents through different resistances.	(iii) the equivalent resistance
4.	(d) The equivalent resistance has a lower value than that of any individual resistance.	(iv) resistances in series
5.	(e) Uninterrupted working of devices in good working condition	

Answers

3. No.

4. Yes.

5. Less than the individual resistances R_1 and R_2.

11. (a) 3.6 Ω (b) 6.67 A (c) 4.0 A, 2.0 A, 0.67 A.

12. 3.75 Ω.

14. (a) 1.13 Ω (b) 5.3 A (c) 0.6 A, 1.5 A, 1.2 A and 2.0 A (d) 6 V.

15. 8.67 Ω.

16. 15 Ω and 10 Ω.

17. (a) 1.0 A, 0.5 A (b) 2.74 A (c) 59.8 V.

18. (a) 2.0 A, 1.0 A (b) 3.0 A.

19. (c)

20. (a)

21. (d)

22. (a)
23. (c)
24. (b)
25. reciprocal, parallel
26. parallel
27. F.
28. T.
29.
(1) [(a), (iii)]
(2) [(b), (iv)]
(3) [(c), (i)]
(4) [(d), (i)]
(5) [(e), (ii)]

Hints/Solutions

11. (a) Let R_p be the equivalent resistance of the parallel combination. Then, $1/R_p = 1/R_1 + 1/R_2 + 1/R_3 = 1/6 + 1/12 + 1/36 = (10/36)$. Or, $R_p = (36/10)\ \Omega = 3.6\ \Omega$.

(b) The total current $I = V/R_p = (24\ \text{V})/(3.6\ \Omega) = 6.67\ \text{A}$.

(c) $I_1 = V/R_1 = (24\ \text{V})/(6\ \Omega) = 4.0\ \text{A}$, $I_2 = V/R_2 = (24\ \text{V})/(12\ \Omega) = 2.0\ \text{A}$, $I_3 = V/R_3 = (24\ \text{V})/(36\ \Omega) = 0.67\ \text{A}$.

Note: We find that $I = I_1 + I_2 + I_3$, which is expected for a parallel combination.

12. The resistor of $5\ \Omega$ and R are in series. So, $5\ \Omega + R = 20\ \Omega$, which gives $R = 15\ \Omega$. The equivalent resistance of the parallel combination is $R_p = (5\ \Omega) \times (15\ \Omega) / (5\ \Omega + 15\ \Omega) = (15/4)\ \Omega = 3.75\ \Omega$.

14. The circuit can be redrawn as shown in Figure 6.19

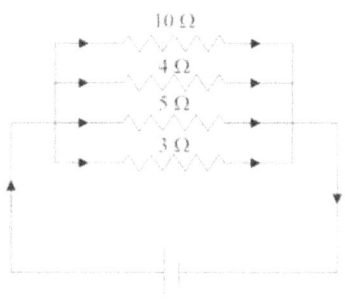

FIGURE 6.19

(a) The equivalent resistance of a parallel combination of four resistors is given by $1/R_p = 1/10 + 1/4 + 1/5 + 1/3 = (53/60)$. Or, $R_p = (60/53)\ \Omega = 1.13\ \Omega$.

(b) The total current $I = V/R_p = (6\ \text{V})/(1.13\ \Omega) = 5.3\ \text{A}$.

(c) $I_1 = V/R_1 = (6\ \text{V})/(10\ \Omega) = 0.6\ \text{A}$, $I_2 = V/R_2 = (6\ \text{V})/(4\ \Omega) = 1.5\ \text{A}$, $I_3 = V/R_3 = (6\ \text{V})/(5\ \Omega) = 1.2\ \text{A}$ (d) $I_4 = V/R_4 = (6\ \text{V})/(3\ \Omega) = 2.0\ \text{A}$

(d) the potential difference across each resistor = 6 V.

Note: We find that $I = I_1 + I_2 + I_3 + I_4$, which is expected for a parallel combination.

15. The resistances of $5\ \Omega$ and $10\ \Omega$ are in parallel. Their equivalent resistance is $R' = (5\ \Omega) \times (10\ \Omega)/(5\ \Omega + 10\ \Omega) = (10/3)\ \Omega$. Again, three resistances of $3\ \Omega$, $4\ \Omega$ and $6\ \Omega$ are in parallel. Let R'' be the equivalent resistance of this parallel combination. Then, $1/R'' = 1/3 + 1/4 + 1/6 = (3/4)\ \Omega$. Or $R'' = (4/3)\ \Omega$.

The circuit can be redrawn as shown in Figure 6.20.

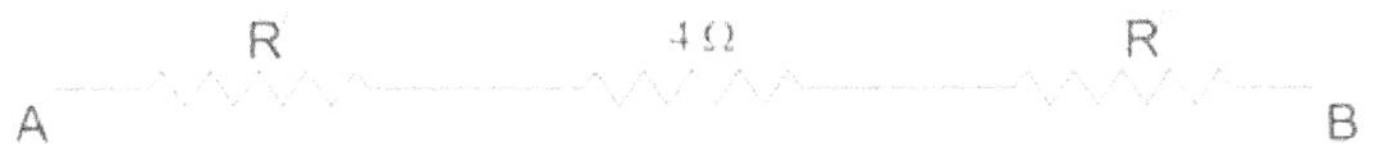

                    R'   4 Ω   R"
            A   ──/\/\/──/\/\/──/\/\/── B

FIGURE 6.20

The three resistances R' and R'' are seen to be joined in series with the third resistance of 4 Ω in the middle. The total resistance between the points A and B, R_T is given by

$R_T = R' + 4\,\Omega + R'' = (10/3)\,\Omega + 4\,\Omega + (4/3)\,\Omega = (26/3)\,\Omega = 8.67\,\Omega$.

16. We have, $R_s = R_1 + R_2 = 25\,\Omega$, $R_p = R_1 \times R_2 / (R_1 + R_2) = 6\,\Omega$.

From above relations, we get $R_2 = 25 - R_1$ and $R_1 \times R_2 / (25) = 6$,

or $R_1 \times (25 - R_1) = 150$, or $25 R_1 - R_1^2 = 150$, which gives the equation

$R_1^2 - 25 R_1 + 150 = 0$, or $(R_1 - 15)(R_1 - 10) = 0$.

Therefore, either, $R_1 = 15$, which gives $R_2 = 25 - R_1 = 10$, or $R_1 = 10$, which gives $R_2 = 15$.

Hence, the two resistances are 15 Ω and 10 Ω.

17. The current through the 4 Ω resistors is 1.25 A.

Using Ohm's law, the potential difference across 4 Ω resistor $= 4\,\Omega \times (1.25\,A) = 5.0\,V$.

(a) We know that in a parallel combination, a common potential difference exists across each of the resistances. So, the potential difference across resistances of 5 Ω and 10 Ω is also 5.0 V.

The current through the 5 Ω resistor $= (5.0\,V/ 5\,\Omega) = 1.0\,A$, and the current through the 10 Ω resistor $= (5.0\,V/ 10\,\Omega) = 0.5\,A$.

(b) The equivalent resistance of a parallel combination of three resistors is given by $1/R_p = 1/4 + 1/5 + 1/10 = (11/20)$. Or, $R_p = (20/11)\,\Omega = 1.82\,\Omega$. The circuit can be redrawn as in Figure 6.21.

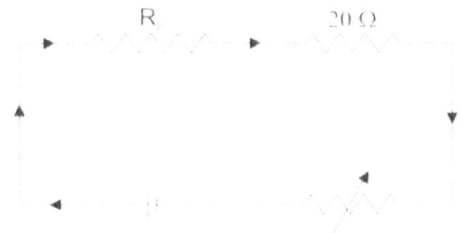

FIGURE 6.21

The potential difference across R_p is 5.0 V. On applying Ohm's law, the current through R_p is $I_p = (5.0\,V/ 1.82\,\Omega) = 2.74\,A$.

Now R_p is in series with 20 Ω resistor. So, the current through 20 Ω resistor is also $I_p = 2.74\,A$).

(c) the equivalent resistance of the series combination is $R_s = R_p + 20\,\Omega = 21.82\,\Omega$. The current through the series combination = 2.74 A. Hence, the reading of the voltmeter $V = (2.74\,A) \times (21.82\,\Omega) = 59.8\,V$.

18. On applying Ohm's law, the current through 4 Ω resistor $= (12.0\,V/ 4\,\Omega) = 3.0\,A$.

(a) Now, the resistances of 5 Ω and 10 Ω are in parallel. Let their equivalent resistance be denoted as R_1. Now, $R_1 = (5\,\Omega) \times (10\,\Omega) / (5\,\Omega + 10\,\Omega) = (10/3)\,\Omega$.

The given circuit is equivalent to the one shown in Figure 6.22. R_1 is in series with 4 Ω resistor.

FIGURE 6.22

So, the current through R_1 is also 3.0 A. The potential difference across R_1 = (3.0 A) × (10/3) Ω = 10.0 V. The same potential difference exists across each of the resistors in parallel combination i.e., across each of the resistances of 5 Ω and 10 Ω. So, the current through 5 Ω resistor = (10.0 V/ 5 Ω) = 2.0 A, and the current through 10 Ω resistor = (10.0 V/ 10 Ω) = 1.0 A (b). The ammeter, being in series with 4 Ω resistor and R_1, also reads 3.0 A.

19. The equivalent resistance of a parallel combination has a lower value than that of any individual resistance.

20. The value of equivalent resistance of a parallel combination is minimum among all combinations.

21. The equivalent resistance of a parallel combination is has a lower value than that of the smallest resistance.

22. In parallel combination, a common potential difference V exists across each of them. Ohm's law gives $I = V/R$ or current is inversely proportional to resistance.

23. The equivalent resistance of a parallel combination has a lower value than that of any individual resistance including the smallest resistance.

24. Smallest resistance can be obtained by combining them in parallel. The equivalent resistance R' is given as $1/R' = 1/R + 1/R + ...\ n\ times = n/R$ or $R' = R/n$.

SUMMARY OF THE CHAPTER

- An arrangement in which two or more resistances are combined together to obtain an effective value different from any of the given resistances is called the combination of resistances.
- An equivalent resistance of a combination of resistances is the value of that single resistor, which can effectively replace the combination without producing any change in the current of potential difference in the circuit.
- Two or more resistances are said to be connected in series when they are connected end to end so that the same current passes through each of them.
- The equivalent resistance of resistances in series equals the sum of the individual resistances.
- The equivalent resistance of a series combination has a higher value than that of any individual resistances.
- Two or more resistances are said to be connected in parallel when they are joined together between two points such that a common potential difference exists across each of them.
- The reciprocal of equivalent resistance of parallel combination equals the sum of the reciprocals of individual resistances.
- The equivalent resistance of a parallel combination has a lower value than that of any individual resistances.
- A parallel circuit has many advantages over the series circuit. These include the simultaneous use of different types of electric devices, uninterrupted working of different devices and higher value of total current through the circuit.

CHAPTER 7

HEATING EFFECT OF ELECTRIC CURRENT

Conversion of electrical energy into other forms of energy

We have learnt that a source of electric energy (like an electric cell or a battery) must continuously supply energy in order to maintain the flow of electric current in a circuit.

The electric energy expended by the source can be converted into other forms of energy. For example, the blades of a an start rotating when electric current flows through it. Moreover, the fan heats up after sometime. In this case, a part of the electric energy supplied by the source appears as mechanical (rotational) energy and the remaining part is dissipated as heat. Similarly, in case of an electric bulb, electric energy is converted into heat energy as well as light energy.

Heating effect of current

When electric current flows through an electric circuit containing only resistors, the electrical energy supplied by the source is converted entirely into heat energy. For example, an electric heater becomes very hot when current passes through its heating element, which is a resistor made of some metallic alloy.

The phenomenon of conversion of electric energy into heat energy when an electric current flows through a resistor is called the heating effect of current.

Expression for heat produced in a resistor-Joule's Law

Suppose a potential difference V is maintained between the two ends X and Y of a resistor of resistance R by joining it to a source of electric energy like a cell or a battery (Figure 7.1). Let the value of the current flowing through the resistor be I. Then the total charge Q flowing across the resistor in time t is given by

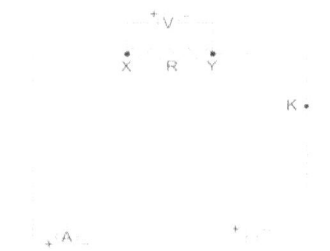

FIGURE 7.1 *Heat produced by a steady current flowing through an electric circuit containing resistor only*

$Q = I\,t$ (7.1)

By definition, the potential difference V is the work done in moving a unit charge from one end of the resistor to the other. So the amount of work done in moving a charge Q is given by

$W = V\,Q$ (7.2)

Using Eq. (7.1) in Eq. (7.2), we get

$W = V\,I\,t$ (7.3)

In order to do this work, the source (i.e. the cell) must be supplying an equal amount of electrical energy to the circuit. Since the circuit contains resistors only, the whole of the electric energy supplied to the circuit gets converted into the heat energy. Therefore, the amount of heat H produced in time t is given as

$H = W = V I t$ (7.4)

By applying Ohm's law to the resistor of resistance R, we get

$V = R I$ (7.5)

Substituting Eq. (7.5) in Eq. (7.4), we get

$H = I^2 R t$ (7.6)

The above relation was first verified experimentally by Joule and is known as 'Joule's law' of heating.

Consequences of Joule's law of heating

Joule's law (Eq. 7.6) shows that

(i) For constant values of R and t,
H is proportional to • I^2
i.e., ***the heat produced in a given resistor in a given time is proportional to the square of the current flowing through the resistor.*** Thus if the current gets doubled, the heat produced will become four times.

(ii) For constant I and t,
H is proportional to R
i.e., ***the heat produced for a given value of current and for a given time interval is directly proportional to the resistance value of the resistor.*** Thus if the same current passes for the same time through two unequal resistances R_1 R_1 and R_2, such that $R_2 = 2 R_1$, the heat produced in R_2 will be double that in R_1.

(iii) I and t being constant,
H is proportional to R
i.e., the heat produced when a given current that flows through a given resistor is directly proportional to the time interval for which the current flows.

Practical Applications of heating effect of current

In many cases the heat produced during passage of current through a resistor can be gainfully utilized for various applications. Some such applications are given below.

(i) Electric heating devices
The electric heater, electric iron, electric oven and toaster are a few well known examples of electric heating devices. All such devices have a heating coil made up of metallic alloys (like nichrome or manganin), which have higher resistivity than pure metals. The heating coil gets heated when electric current flows through it. The Joule heating thus produced finds a specific application in case of these devices.

(ii) Electric bulb

An electric bulb is a device in which Joule heating is used to produce light. The bulb is fitted with very thin filament having high resistance (Figure 7.2), which gets heated when an electric current passes through it. The filament retains most of the heat produced and becomes so hot as to start emitting light. It is essential that the heat produced does not melt the filament. Tungsten is the preferred material for fabrication of the filament because it has very high melting point (3380^0 C).

The bulb is thermally insulated by using proper insulating support. Moreover the bulb is filled with chemically inactive gases like argon or nitrogen to prolong its life.

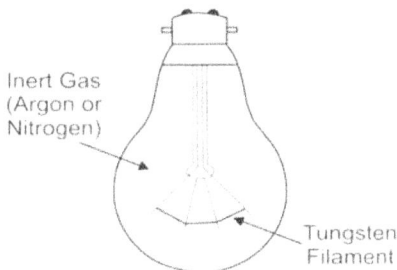

FIGURE 7.2 *The filament of electric bulb emits light when current passes through it.*

It is worth noting that only a small part of the total electric energy consumed by the filament is converted into light energy-the rest being dissipated as heat. Consequently, an electric bulb is not a very efficient device to produce light.

(iii) The electric fuse

An electric fuse is an important safety device used to protect an electric device or an electric circuit by preventing the flow of excessively large a current through it. Its working is based on the heating effect of current. An electric fuse is essentially a small piece of wire generally made up of an alloy of tin and copper having low melting point. One type fuse wire used in domestic electric circuits, is generally fitted in a cartridge of some insulation material like porcelain (Figure 7.3).

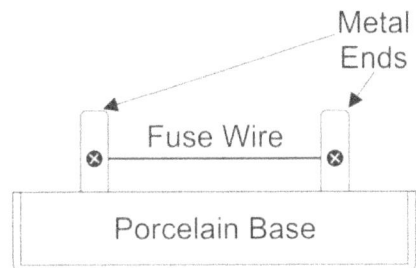

FIGURE 7.3 *The fuse wire melts due to Joule heating when the current exceeds the safe value and the circuit is broken.*

The fuse is placed in series with the device to be protected so that the same current passes through it as through the device. When this current exceeds a certain safe value, due to Joule heating, the temperature of the fuse wire increases so much that the latter melts and the circuit is broken so that current does not flow any more in the circuit. In this way, the fuse wire prevents any damage to the device or the circuit.

Disadvantages of heating effect of current

Heating effect of current has both advantages and disadvantages. Having listed above some of the former, let us now mention two of its major disadvantages.

In many cases, when current passes through resistor, dissipation of energy in the form of heat results in the wastage of energy. This becomes particularly disadvantageous during flow of current in transmission lines over long distances from power plants to the places of utilization. Similarly, in electric motors (used in many devices such as fans, mixers, washing machines etc.), a part of the input electric energy gets dissipated as wasteful heat.

The unavoidable heat produced also results in the rise in the temperature of the components through which the current flows. Higher temperature produces undesirable changes in the properties (including resistance values) of these components. The changed values might adversely affect the reliable functioning of the electric circuit containing these components.

SOLVED EXAMPLES

7.1 Heating Effect of Current

When electric current flows through a metal wire of resistance 6 Ω, the heat produced in 8 s is 1200 J. Find the value of the current and the potential difference across the resistor.

Solution

We know that the heat produced H, when a current I flows through a resistance R for time t is

$H = I^2 R t$, which gives

$I^2 = H/RT$

Or, $I = \sqrt{(H/RT)}$

Here, H = 1200 J, R = 6 Ω, t = 8 s.

So, $I = \sqrt{(H/R t)} = [\sqrt{\{(1200 \text{ J})/(6 \text{ Ω} \times 8 \text{ s})\}}]$ A = $\sqrt{25}$ = 5 A.

To find the potential difference V, use Ohm's law,

$V = I R$ = (6 Ω) (5 A) = 30 V.

7.2 Heating Effect of Current

An electric heater of resistance 30 Ω is connected across the line voltage of 220 V. Calculate the heat develop in 1 minute.

Solution

Resistance, R = 30 Ω, line voltage V = 220 V

On applying Ohm's law, we get

current, $I = V/R$ = 220 V/ 30 Ω = (22/3) A.

Time, t = 1 minute = 60 s.

So, heat developed, $H = I^2 R t$

= (22/3 A)^2 × (30 Ω) (60 s) = 96,800 J.

7.3 Heating Effect of Current

A bulb is connected across a 12 V battery; the resistance of the bulb filament is 4 Ω. Calculate the energy transferred to the bulb in 10 minutes.

Solution

Battery voltage, V = 12 V.

Resistance of the bulb, R = 4 Ω.

Time, t = 10 min = 600 s.

Current flowing through filament, I is obtained by applying Ohm's law.

$I = V/R$ = 12 V/4 Ω = 3 A.

Energy transferred to the bulb in time t is = $I^2 R t$.

Using the above values, energy transferred = $(3\ A)^2 \times (4\ \Omega) \times (600\ s) = 21{,}600\ J$.

7.4 Heating Effect of Current

Compute the heat generated when a current of 10 A passes through a heater coil for 1hr when connected across the line voltage of 220 V.

Solution

Current $I = 10$ A.

Potential difference, $V = 220$ V.

Time, $t = 1\text{hr} = 3600$ s.

We know that the heat produced in time t when current I flows through a potential difference V is

$H = V I t$

So, $H = (220\ V) \times (10\ A) \times (3600\ s)$

$= 7.92 \times 10^6$ J.

EXERCISES E 7.1

Based on Heating Effect of Electric Current

A. Very Short Answer Type Questions

1. Name the SI unit of electric energy?
2. Define electrical energy. Name one source of electrical energy.
3. Give the mathematical formula representing the Joule's law of heating.
4. Name four electric devices making use of heating effect of current.
5. How much electric energy is supplied by a source of potential difference V, which maintains a steady current I through a circuit for time t?
6. Name the material commonly used for making fuse wire.

B. Short Answer Type Questions – I

7. *(Numerical Problem)* How much heat is produced in a conductor when a charge of 4800 C passes through it in half an hour if the potential difference across the conductor is 24 V?
8. *(Numerical Problem)* A battery sends a current of 3 A through a 5 Ω resistor. Find the electric energy supplied by the battery in 15 minutes.

C. Short Answer Type Questions – II

9. *(Based on Higher Order Thinking Skills (HOTS))* How will the electric energy consumed in time t by an electric heater operating at a potential V be affected if the length of its heating element decreases?
10. *(Numerical Problem Based on Higher Order Thinking Skills (HOTS))* We have two resistors R_1 and R_2 having resistance values 5 Ω and 10 Ω respectively. Find the ratio of the heat produced in R_1 and R_2 in 5 minutes, when they are joined in series across a battery of 5 V.
11. *(Numerical Problem Based on Higher Order Thinking Skills (HOTS))* Two metallic wires having resistances $R_1 = 5\ \Omega$ and $R_2 = 10\ \Omega$ are joined in parallel across a battery. What will be the ratio of heat produced by R_1 and R_2 in time t?
12. *(Numerical Problem Based on Higher Order Thinking Skills (HOTS))* An electric bulb consumes an electric energy of 100 J in 5 minutes, when a current I passes through its filament. How much energy will it consume in 10 minute if the current through its filament is $2I$?

D. Long Answer Type Questions

13. Derive the expression for the heat produced in a resistance *R* through which a current *I* flows for time *t*. What is this expression known as?
14. Discuss three application of the heating effect of current.
15. What is a fuse? How does it protect an electric circuit?
16. State Joule's law of heating? What are its implications?
17. How does the heat produced in a resistance *R* depend on (i) the current (ii) the value of *R* and (iii) time for which current flows? Name the law which governs this dependence.
18. How is the heating effect of current used in the working of an electric bulb? What is the filament of the bulb made of and why?

E. Objective Questions

I. Multiple Choice Type Questions

Choose the Correct Answer:

19. The electric cell is a device to obtain
(a) electrons
(b) electric energy from chemical energy
(c) electric charge
(d) electric force.
20. Which of the following is converted into heat when current is set up through a conductor?
(a) electric current
(b) electric potential
(c) electric energy
(d) electric resistance.
21. Tungsten is used for fabrication of the filament of an electric bulb. This is because
(a) its melting point is very low
(b) its melting point is very high
(c) its resistivity is low
(d) its resistance is independent of temperature
22. The fuse is placed
(a) in parallel with the device to be protected
(b) anywhere in the circuit
(c) in series with the device to be protected
(d) none of the above
23. The metallic alloy used for fabrication of a fuse
(a) should have high melting point
(b) should have high resistivity
(c) should have high resistance
(d) should have low melting point
24. The amount of heat H produced in a resistor *R* in time t is given as
(a) $H = I R t$
(b) $H = I R^2 t$
(c) $H = I^2 R t$
(d) none of the above
25. *(Based on Higher Order Thinking Skills (HOTS))* The amount of heat produced in ten minutes a given circuit
(a) is more in a thick copper wire than a thin copper wire of same length
(b) is less in a thick copper wire than a thin copper wire of same length
(c) can either be more or less depending on room temperature
(d) none of the above

II. Fill in the blanks Type

Fill in the blanks:

26. The heat produced in a given resistor in a given time is proportional to the …….. of the current flowing through the resistor.
27. The electric bulb is fitted with very ……. filament having …… resistance
28. Tungsten is the preferred material for fabrication of the filament of a bulb because it has very ….. melting point.
29. The fuse is placed in …… with the device to be protected so that the same …… passes through it as through the device.

III. True or False Type

Mark the following Statements True (T) or False (F)

30. The heat produced for a given value of current and for a given time interval is directly proportional to the resistance value of the resistor. T/F
31. An electric bulb has a heating coil made up of a metallic alloy. T/F
32. The working of a fuse is based on the heating effects of current. T/F

Answers

5. Electric energy supplied = VIt.	12. 800 J.	26. square
6. An alloy of tin and copper.	19. (b)	27. thin, high
7. 115,200 J.	20. (c)	28. high
8. 40,500 J.	21. (b)	29. series, current
9. The energy consumed will increase.	22. (c)	30. T.
	23. (d)	31. F.
10. 1:2.	24. (c)	32. T.
11. 2:1.	25. (b)	

Hints/Solutions

7. The amount of heat produced in time t when current I flows through a potential difference V is $H = VIt$. If Q be the charge flowing in time t, we have $I = Q/t$. So, $H = V \times (Q/t) \times t = VQ = (4800\ C) \times (24\ V) = 115{,}200$ J.

8. $I = 3$ A, $R = 5\ \Omega$, $t = 15$ min $= 900$ s. We know that supplied by the battery in time t is $= I^2 R t$.
Using the above values, energy supplied $= (3\ A)^2 \times (5\ \Omega) \times (900\ s) = 40{,}500$ J.

9. The electric energy consumed in time t when current I flows through a potential difference V is $= VIt$. If R be the resistance of the heating element, $I = V/R$, the energy consumed $= V \times (V/R) \times t = (V^2/R) \times t$.
The above result shows that for given V and t, energy consumed is inversely proportional to resistance R. Now, decrease in length implies decrease in R ($= \rho\, l/A$). So, the electric energy consumed in time t by an electric heater operating at a potential V will increase if the length of its heating element decreases.

10. We know that in a series circuit, a common current I flows through each of the resistors. I flows for same time t (= 5 minutes) in both of them.
So, the ratio of the heat produced in R_1 and R_2 in 5 minutes $= (I^2 R_1 t)/(I^2 R_2 t) = R_1/R_2 = (5\ \Omega)/(10\ \Omega) = 1{:}2$.

11. We know that in a parallel combination, a common potential difference V exists across each of the resistances in parallel.
So, currents through R_1 and R_2 are $I_1 = V/R_1$ and $I_2 = V/R_2$.
Ratio of the heat produced in R_1 and R_2 in time $t = (I_1^2 R_1 t)/(I_2^2 R_2 t)$

= {$(V/R_1)^2 \times R_1$}/ {$(V/R_2)^2 \times R_2$} = R_2/R_1 = (10 Ω)/ (5 Ω) = 2:1.

12. When current I flows for 5 minutes (= 300 s), energy U consumed is $U = I^2 \times R \times (300\ s)$ = 100 J (Given). So, when current ($2I$) flows for 10 minutes (= 600 s) through the same filament, energy consumed will be $U' = (2I)^2 \times R \times (600\ s)$ = 8 × U = 8 × (100 J) = 800 J.

19. In an electric cell or a battery, the electrical energy is obtained from the chemical energy of the cell.

20. When electric current flows through an electric circuit containing resistors only, the electrical energy supplied by the cell is converted into heat energy.

21. Tungsten is used for fabrication of the filament of electric lamps because it has very high melting point (3380^0 C).

22. The fuse is placed in series with the device to be protected so that the same current passes through it as through the device.

23. A fuse is made up of an alloy (such as an alloy of tin and copper), which has low melting point.

24. Joule's law gives the amount of heat H produced in a resistor R in time t as $H = I^2 R t$.

25. According to Joule's law, for constant I and t, H is proportional to R. Heat produced is less because a thick copper wire has a lower resistance than a thin copper wire of same length.

SUMMARY OF THE CHAPTER

- Electrical energy is that form of energy which is required for the flow of current in an electric circuit.
- The phenomenon of conversion of electric energy into heat energy when an electric current flows through a resistor is called the heating effect of current.
- Joule's law tells us that the heat produced in a resistor of resistance R when a current I passes through it for time t is $H = I^2 R t$.

  In other words, (i) for constant R and t, H is proportional to the square of the current I (ii) for constant I and t, H is proportional to the resistance R, and (iii) for constant R and I, H is proportional to the time interval t for which the current flows.
- Practical applications of heating effect of current include the electric heating devices (like the electric room heater, electric iron, electric oven and toaster), the electric bulb and the electric fuse.
- An electric bulb is a devise in which Joule heating of a filament is used to produce light.
- The electric fuse is a safety device, which protects an electric device or an electric circuit by preventing the flow of too large a current through it.
- One disadvantage of heating effect of current is that the dissipation of energy in the form of heat results in the wastage of energy. Another disadvantage is that heat produced results in the changes in the properties (including resistance values) of these components. The changed values might adversely affect the reliable functioning of the electric circuit containing these components.

CHAPTER 8

ELECTRIC POWER AND COMMERCIAL UNIT OF ELECTRIC ENERGY

Definition of electric power

We know that power is defined as the rate of doing work or the rate at which energy is consumed. In case of electric energy, *we can define electric power as the rate at which electrical energy is consumed or dissipated in the electric circuit.*

Suppose a source (a cell or a battery) maintains a potential difference V across an electric circuit through which a steady current I flows for time t. Then the amount of work done in moving a charge Q (= $I\,t$) is given by

$W = V\,Q$

$= V\,I\,t \qquad (8.1)$

So, the energy supplied by the source, or equivalently, the energy U consumed (or dissipated) in time t also equals

$U = W = V\,I\,t \qquad (8.2)$

By definition, the electric power

$P = U/t = V\,I \qquad (8.3)$

When the circuit contains resistance R, then by Ohm's law, $V = I\,R$. Substituting the value of V in Eq. (8.3), we get

$P = I^2\,R \qquad (8.4)$

or equivalently, using $I = V/R$, we have

$P = V^2/R \qquad (8.5)$

Unit of electric power

The SI unit of electric power is watt, which is named after the English scientist James Watt (1736- 1819). From Eq. (8.3), we have

1 watt = 1 volt × 1 ampere

or $1\ W = 1\ V\,A \qquad (8.6)$

We may, therefore, define one watt as the power consumed by a circuit when a current of 1 A passes through it at a potential difference of 1 V.

Watt is a very small unit of electric power. For practical purposes, we make use of a higher unit, the kilowatt (kW). We have 1 kW =1000 W.

Commercial unit of energy-kilowatt hour

We know that electric energy = electric power × time. When electric power is expressed in kilowatt and time in hours, the unit of electrical energy is 1 kilowatt hour (denoted by kW h). ***One kilowatt hour is the energy consumed when 1 kilowatt of power is used for one hour.***

Kilowatt hour is known as commercial unit of electric energy because the cost of commercially available electricity is measured in terms of kilowatt hour. It is commonly known as the 'unit' of electricity. We can express 1 kW h in terms of joules as follows.

1kW h

= 1 kW × 1 hour

= 1000 watt × 3600 second

= 3.6 × 10^6 watt sec

But, *1 watt sec = 1 joule (J)*

So, *1 kW h = 3.6 x 10^6 J*
$$(8.7)$$

Rating of electric appliances

Electric appliances like electric bulb, fan, toaster, geysers etc. have certain specification written on them, which mentions values of voltage and power. The voltage value corresponds to the safe value of potential difference that can be applied across the input terminals of the appliance. The power value (or the wattage) refers to the rate of consumption of electric energy when the appliance is being used at the rated voltage. The pair of voltage and power values is known as the rating of the appliance.

As an example, consider an electric heater having a rating of (220 V, 1000 W). This heater consumes electric energy at the rate of 1000 J/s, when it is used across the line voltage of 220 V. If it is used continuously for 1 hour, the electric energy consumed by the heater can be calculated by using the formula, electric energy = power × time = 1000 W × 1 h = 1000 W h = 1 kW h. The heater would, thus, consume one unit of electric energy in one hour of continuous use.

Knowing its rating, we can easily calculate the current flowing through the appliance by making use of the formula, $P = V \times I$, which gives $I = P/V$. Thus, a (220 V, 1000 W) heater draws a current of I = 1000 W/220 V = 4.55 A.

The rating of a heating device also allows us to know the resistance R of the heating element of the devise. Using the formula $P = V^2/R$, we get $R = V^2/P$. In the above example of electric heater with rating (220 V, 1000 W), the heating element would have a resistance of $R = ((220)^2/1000)$ Ω = 48.4 Ω.

Choice of an appropriate fuse

Fuses for different ratings such as 1 A, 3 A, 5 A, 10 A, 15 A and higher values are available in the market.

One can have an idea of the fuse that should be used for protecting a particular appliance by knowing the latter's rating. Suppose an appliance has a rating of (220 V, 2 kW). The current I flowing through the appliance, when used across a line voltage of 220 V is 9.1 A ($I = P/V$ = 2 kW/220 V). For a safe functioning of this appliance, a fuse of 10 A rating should be put in series with the appliance. If, due to voltage fluctuation or some other reason, the current exceeds 10 A, the fuse wire would melt. This results in the breaking of the circuit and the appliance is prevented from getting damaged.

Solved Examples

8.1 Electric Power used in Resistors

An electric bulb joined across a 220 V supply wire has a current of 0.75 A flowing through its filament. What is the power of the bulb?

Solution

We have, $V = 220$ V, $I = 0.75$ A.

The power P if the bulb is given as

$P = IV = (0.75$ A$) \times (220$ V$) = 165$ W.

8.2 Electric Energy and Power

(Based on Higher Order Thinking Skills (HOTS))

An electric iron connected across a voltage of 220 V consumes energy at the rate of 1000 W when heating is at the maximum rate and 400 W when heating is at minimum rate. What are the values of the current and the resistance in each case?

Solution

We know that the power P consumed by an appliance when a current I flows through it at potential difference V is

$P = VI$,

Or, $I = P/V$.

(i) When heating is at the maximum rate,

$P = 1000$ W, $V = 220$ V.

So, $I = P/V = (1000$ W$)/ (220$ V$) = 4.55$ A.

Using Ohm's law, we get resistance

$R = V/I = (220$ V$)/ (4.55$ A$) = 48.4$ Ω.

(ii) When heating is at the minimum rate,

$P = 400$ W, $V = 220$ V.

So, $I = P/V = (400$ W$)/ (220$ V$) = 1.82$ A.

Again, using Ohm's law, we get

$R = V/I = (220$ V$)/ (1.82$ A$) = 120.88$ Ω.

8.3 Electric Energy and Power

(Based on Higher Order Thinking Skills (HOTS))

How many electric bulbs each of rating 15 W - 220 V can be connected in parallel with each other across the two wires of 220 V line if the maximum allowable current is 12 A?

Solution

Let the desired number of bulbs = n

Maximum allowable current = 12 A

Power of each bulb $P = 15$ W

Voltage rating of each bulb $V = 220$ V

Current drawn by each bulb $I = P/V = 15$ W/220 V = (3/44) A

We know that a common potential difference V exists across each of the resistances in parallel. So the potential difference across each of the bulbs is also 220 V. Each bulb, therefore, continues to draw a current of (3/44) A.

Now, the total current flowing in a parallel combination equals the sum of separate currents through different branches. Here, number of branches = n and the current in each branch = I.

Clearly,

$n \times I = 12$ A

or $n = 12$ A$/ I = 12$ A$/ ((3/44)$ A$) = 176$.

8.4 Electric Energy and Power

How many commercial units of electric energy are consumed per hour by an electric heater which is drawing 4 A from a 220 V line?

Solution

Current through the motor, $I = 4$ A.

Line voltage, $V = 220$ V, Time, $t = 1$ h.

So the power of motor is

$P = V I = (220$ V$) \times (4$ A$) = 880$ W.

Energy consumed in 1 h = $P t$

= (880 W) × (1 h)

= 880 W h = 2.20 kW h = 2.20 units.

8.5 Electric Energy and Power

What is the difference in cost of electric energy consumed by a 2000 W electric geyser in 1 hr and a 200 W TV set in 8 hours if the electricity costs *Rs.* 4.0 per unit?

Solution

We know that electric energy = electric power × time

So the electric energy used by electric geyser

= 2000 W × 1 h = 2000 W h.

And electric energy used by TV set

= 200 W × 8 h = 1600 W h.

The difference of electric energy = 2000 W h – 1600 W h

= 400 W h = 0.4 kW h = 0.4 units.

So, the difference in cost of electric energy = 0.4 units × Rs. 4.0/ unit

= Rs. 1.60.

8.6 Electric Power and Energy

An electric iron of resistance 5 Ω draws 10 A from the service mains for 30 minutes. Calculate the rate at which heat is developed in the iron.

Solution

Current drawn from the mains $I = 10$ A

Resistance of the electric heater $R = 5\ \Omega$

We know that the rate at which heat is developed i.e. power P is given as

$P = I^2 R = (10\ A)^2 \times (5\ \Omega) = 500$ W.

8.7 Electric Power and Energy

A 60 W electric bulb is used for two hours and a 40 W tube light is used for 4 hours per day on an average. What is the cost of running these devices for a month of 30 days when the electricity costs *Rs.* 3.0 per unit?

Solution

The electric energy consumed /day for running electric bulb

$= 60\ W \times 2\ h = 120\ W\ h$.

The electric energy consumed per day for running the tube light

$= 40\ W \times 4\ h = 160\ W\ h$.

So, the total energy consumed per day

$= 120\ W\ h + 160\ W\ h$.

$= 280\ W\ h = 0.28\ kW\ h$.

The total energy consumed per month

$= 30$ days $\times\ 0.28$ kW h/day $= 8.40$ kW h $= 8.40$ units.

Hence the cost of electric energy consumed

$= 8.40$ units $\times$ Rs. 3.0/ unit = Rs. 25.20.

8.8 Electric Power

Two lamps, one rated 220 V - 60 W and the other 220 V - 40 W, are connected in parallel to electric mains supply. What current is drawn from the line if the supply voltage is 110 V?

Solution

Line voltage $V = 110$ V.

The formula $P = V^2/R$ gives

$R = V^2 / P$

Resistance of the first lamp $R_1 = V^2 / P_1 = (220\ V)^2 /60\ W = 806.7\ \Omega$

Resistance of the second lamp $R_2 = V^2 / P_2 = (220\ V)^2 /40\ W = 1210.0\ \Omega$.

Since the lamps are in parallel, their equivalent resistance is

$R_p = R_1 \times R_2 / (R_1 + R_2) = (806.7\ \Omega \times 1210.0\ \Omega)/ (806.7\ \Omega + 1210.0\ \Omega) = 484.01\ \Omega$

So, current drawn from the line $= V/R_p = 110$ V/484.01 Ω = 0.23 A.

8.9 Electric Power used in Resistors

(Based on Higher Order Thinking Skills (HOTS))

Two bulbs are rated 220 V - 100 W and 220 V - 50 W respectively. Which bulb will grow brighter if they are connected

(a) in parallel

(b) in series to 220 V mains? Find also the heat generated per unit time by each bulb in cases (a) and (b)

Solution

Supply voltage $V = 220$ V

Power of the first lamp $P_1 = 100$ W

Power of the second lamp $P_2 = 50$ W

Resistance of the first lamp $R_1 = V^2/P_1 = (220 \text{ V})^2/100 \text{ W} = 484 \, \Omega$

Resistance of the second lamp $R_2 = V^2/P_2 = (220 \text{ V})^2/50 \text{ W} = 968 \, \Omega$.

(a) We know that in parallel combination, a common potential difference V exists across each of the bulbs.

Heat generated per unit time i. e., power consumed by first bulb = $V^2/R_1 = 100$ W.

Power consumed by second bulb = $V^2/R_2 = 50$ W.

Since 220 V - 100 W is consuming more power, it will glow brighter in parallel combination.

(b) We know that in series combination, a common current I flows through each of the bulbs.

The total resistance in the series circuit = $R_1 + R_2$

So, common current $I = V/(R_1 + R_2) = 220 \text{ V}/(484 + 968) \, \Omega = 0.15$ A.

Heat generated per unit time i. e., power consumed by first bulb = $I^2 R_1 = (0.15 \text{ A})^2 \times 484 \, \Omega = 10.9$ W.

Power consumed by second bulb = $I^2 R_2 = (0.15 \text{ A})^2 \times 968 \, \Omega = 21.8$ W.

So, in series combination, it is 220 V - 50 W bulb, which is consuming more power and hence, will glow brighter.

8.10 Electric Power and Energy

An electric heater is rated 1000 W, 220 V. What is the resistance of its coil? If the voltage drops to 192 V, how much power will it consume and what will be current through its coil?

Solution

Power, $P = 1000$ W, $V = 220$ V.

Let R be the resistance of the heater coil, we have $P = V^2/R$, which gives

$R = V^2/P$

$= (220 \text{ V})^2/1000 \text{ W} = 48.4 \, \Omega$.

Resistance remains unchanged when the voltage changes to $V' = 192$ V. So the power consumed will be

$P = (V')^2/R = (192)^2/48.4 \, \Omega = 761.65$ W.

The current through the coil, $I = V'/R = 192 \text{ V}/48.4 \, \Omega = 3.97$ A.

8.11 Electric Power and Energy

(Based on Higher Order Thinking Skills (HOTS))

Two heating elements, each having resistance R are available to us for heating water. Should they be joined in series or parallel with the mains for faster heating?

Solution

Let the supply voltage be V.

When the heating elements are joined in series, the equivalent resistance is $R_s = R + R = 2R$.

The power consumed by the series combination $P_1 = V^2/R_s = V^2/2R$.

When they are joined in parallel, the equivalent resistance is $R_p = R \times R/(R + R) = R/2$.

The power consumed by the parallel combination $P_2 = V^2/R_p = V^2/(R/2) = 4 \times (V^2/2R) = 4 \times P_1$.

Since more power is consumed in parallel combination, the heating elements should be joined in parallel with the mains for faster heating.

EXERCISES E 8.1

Based on Electric Power

A. Very Short Answer Type Questions

1. Give the formula of electric power in terms the resistance R of a conductor and the potential difference V across it.
2. Name the SI unit of electric power.
3. Is kilowatt hour (kW h) a unit of electric power?
4. Name the commercial unit of electric energy.
5. How much power is consumed by a circuit through which a current I flows when the potential difference across it is V?

B. Short Answer Type Questions – I

6. What determines the rate at which energy is consumed in an electric circuit having a resistance R, when a current I flows through it?
7. What is the commercial unit of electric energy? Express its value in terms of SI unit of energy.
8. *(Numerical Problem)* A torch bulb works with a cell of 2.5 V, and the current through is measured to be 500 mA. Find (i) the power consumed and (ii) the resistance of the bulb.
9. *(Numerical Problem)* An electric bulb joined across a 220 V supply wire has a current of 0.75 A flowing through its filament. What is the power of the bulb?

C. Short Answer Type Questions – II

10. How much power P is consumed by a device carrying a current I when operated at a potential difference V? Also write the expression for P if the device operating at the same voltage is a resistor having a resistance R?
11. *(Numerical Problem)* An electric heater of power rating of 500 W operates 4 hour/day. What is the cost of electricity to run it for 30 days at a rate of Rs. 3.50 per kW h?

D. Long Answer Type Questions

12. What is electric power? Obtain the formula for electric power P in terms of
(i) the current I through circuit, when the potential difference across it is V.
(ii) the resistance R of a conductor having a current I flowing through it?
(iii) the resistance R of a conductor having a potential difference V across it.
13. What is the SI unit of electric power and how is it defined? What is a commercial 'unit' of electric energy known as? How the latter related to Joule?
14. *(Numerical Problem Based on Higher Order Thinking Skills (HOTS))* Two electric lamps of 100 W and 25 W respectively are joined in parallel to a supply of 200 V. Calculate (a) the total current flowing through the circuit (b) the total power consumed by the lamps.
15. *(Numerical Problem Based on Higher Order Thinking Skills (HOTS))* A 100 W electric bulb is used for two hours and a 60 W electric fan is used for 8 hours per day on an average. What is the cost of running these devices for a month of 30 days when the electricity costs Rs. 4.0 per unit?

E. Objective Questions

I. Multiple Choice Type Questions

Choose the Correct Answer:

16. A current of I ampere is flowing through a resistor of R ohm. The rate of heat generated in joules will be:
(a) I^2R
(b) IR^2
(c) IR
(d) given by none of the above expressions.

17. A heater is marked 1000 W. The energy consumed by it in one hour is
(a) 1 J
(b) 1 kJ
(c) 1 kW h
(d) 1 W h.

18. Resistance of conductor is doubled keeping the potential difference across it constant. The rate of generation of heat will:
(a) become one fourth
(b) be halved
(c) be doubled
(d) become four times.

II. Fill in the blanks Type

Fill in the blanks:

19. .... is known as commercial unit of electric energy.

III. True or False Type

Mark the following Statements True (T) or False (F)

20. One kW h = 3.6×10^3 kilo joule. T/F

21. An electric heater having a rating of (220 V, 2000 W) consumes one unit of electric energy in one hour of continuous use. T/F

IV. Matching Type Questions

22. Select the pairs, in the two columns, that match each other

S. No.	Column A	Column B
1.	A_1 Joule's Law	B_1 the filament made of Tungsten
2.	A_2 The electric bulb	B_2 unit of electric energy
3.	A_3 An electric fuse	B_3 unit of electric power
4.	A_4 electric power	B_4 Heating effect of current
5.	A_5 Kilowatt hour	B_5 a safety device
6.	A_6 kilowatt	B_6 the rate at which electrical energy is consumed

V. Crossword Puzzle

23. Complete the crossword puzzle with the help of given clues

The Clues
Across:
1. Commercial unit of electric energy
2. A device which makes use of Joule heating to produce light

Down:
3. SI unit of energy
4. The working of an electric fuse is based on the .... effect of current.
5. An important safety device used to protect an electric circuit by preventing the flow of excessively large a current through it.

Answers

3. No.
5. Electric power $P = V \times I$.
8. (i) 1.25 W (ii) 5 Ω
9. 165 W.
11. Rs 210.00.
14. (a) 0.625 A (b) 125 W.
15. Rs. 81.60.
16. (a)
17. (c)
18. (b)
19. Kilowatt hour
20. T.
21. F.
22.
(1) [A_1; B_4]
(2) [A_2; B_1]
(3) [A_3; B_5]
(4) [A_4; B_6]
(5) [A_5; B_2]
(6) [A_6; B_3]
23.
Across:
1. kilowatt hour
2. bulb
Down:
3. joule
4. heating
5. fuse

Hints/Solutions

8. Here $V = 2.5$ V, $I = 500$ mA $= 500 \times 10^{-3}$ A $= 0.5$A

(i) The power consumed $P = VI = 2.5$ V $\times 0.5$ A $= 1.25$ W

(ii) Using Ohm's law, we get,

resistance of the bulb $R = V/I = (2.5 \text{ V}) / (0.5 \text{ A}) = 5 \, \Omega$.

9. We have, $V = 220$ V, $I = 0.75$ A.

The power P of the bulb is $P = V \times I = (220 \text{ V}) \times (0.75 \text{ A}) = 165$ W.

11. The electric energy used by per day by the electric heater $= P \times t = 500 \text{ W} \times 4 \text{ h} = 2000$ W h $= 2$ kW h.

So, the total energy consumed in 30 days $= 30 \times 2$ kW h $= 60$ kW h. The cost of electric energy consumed $= 60$ kW h $\times$ Rs. 3.50/ kW h = Rs. 210.0.

14. Since the two lamps are joined in parallel, the potential difference V across each = 200 V.

(a) Currents drawn by 100 W lamp and 25 W lamp are, respectively, $I_1 = P_1/V = (100 \text{ W})/(200 \text{ V}) = 0.50$ A, $I_2 = P_2/V = (25 \text{ W})/(200 \text{V}) = 0.125$ A.

For a parallel combination, the total current $I = I_1 + I_2 = 0.625$ A

(b) the total power consumed by the lamps $P = V \times I = V \times (I_1 + I_2) = V I_1 + V I_2 = P_1 + P_2 = 100$ W $+ 25$ W $= 125$ W.

15. The electric energy consumed /day for running electric bulb

$= 100$ W $\times 2$ h $= 200$ W h.

The electric energy consumed per day for running the fan $= 60$ W $\times 8$ h $= 480$ W h.

So, the total energy consumed per day

$= 200$ W h $+ 480$ W h. $= 680$ W h $= 0.68$ kW h. The total energy consumed per month $= 30$ days $\times 0.68$ kW h/day$=20.40$ kW h $= 20.40$ units. Hence the cost of electric energy consumed $= 20.40$ units $\times$ Rs. 4.0/ unit = Rs. 81.60.

16. According to Joule's law, $H = I^2 R t$. So, rate of heat generated $= H/t = I^2 R$.

17. Energy consumed $= 1000$ W $\times 1$ h $= 1000$ W h $= 1$ kWh.

18. Use of $P = V^2/R$ shows that for constant V, P is proportional to $1/R$. So, doubling of R leads to half value of P.

SUMMARY OF THE CHAPTER

- Electric power P is the rate at which electrical energy is being generated by the source or getting consumed in an electric circuit.
- The SI unit of electric power is watt, which is the energy consumed by circuit when a current of 1 A passes through it at a potential difference of 1 V.
- When a source maintains a potential difference V across an electric circuit through which a steady current I flows
  $P = V I$.

  When the circuit contains resistance R

  $P = I^2 R$ or $P = V^2/R$
- Kilowatt hour (kW h) is known as commercial unit of electric energy because the cost of commercially available electricity is measured in terms of kilowatt hour.
- 1 kW h = 3.6 x 10^6 joule.

QUESTION BANK

1. What does an electric circuit mean?

Answer
An electric circuit is a closed and continuous path in which an electric current can flow.

2. Define the unit of current

Answer
The SI unit of current, known as ampere, is that much current which flows through a cross-section of a conductor when one coulomb of charge flows through it in one second.

3. Calculate the number of electrons constituting one coulomb of charge?

Answer
Charge of 1 electron, $e = 1.6 \times 10^{-19}$ C.
Let the number of electrons in one coulomb of charge = n.
So, total charge = $n \times e = 1$ C.
Or $n = 1$ C$/e = 1$ C $/ (1.6 \times 10^{-19}$ C$) = 6.25 \times 10^{18}$.

4. A current of 1 A is drawn by a filament of an electric bulb. Number of electrons passing through a cross section of the filament in 16 seconds would be roughly
(a) 10^{20}
(b) 10^{16}
(c) 10^{18}
(d) 10^{23}

Answer
(a)
(**Hint:** Charge on one electron, $e = 1.6 \times 10^{-19}$ C, current, $I = 1$ A. By definition of ampere, 1 C of charge flows per second. Charge, $Q = I \times t = (1.0$ A$) \times (16$ s$) = 16$ C. Number of electrons = $Q/e = 16$ C$/ (1.6 \times 10^{-19}$ C$) = 10^{20}$.)

5. State and define the unit of current.

Answer
SI unit of current is ampere. One ampere is that much current which flows through a cross-section of a conductor when one coulomb of charge passes through it in one second.

6. Name a device that helps to maintain potential difference across a conductor

Answer
An electric cell

7. How is a voltmeter connected in the circuit to measure the potential difference between two points?

Answer
A voltmeter is connected in parallel across the two points to measure the potential difference between them.

8. What is meant by saying that potential difference between two points is 1 V?

Answer

The potential difference between two points is one volt when one joule of work is done in moving a charge of one coulomb from one point to the other.

9. What is meant by saying that the potential difference between two points is 1 volt? Name a device that helps to measure the potential difference across a conductor.

Answer
The potential difference between two points is 1 volt when one joule of work is done to move a charge of one coulomb from one point to the other.
The device is called voltmeter.

10. Which of the following represents voltage?
(a) Work done/ (current × time)
(b) Work done Charge
(c) (Work done × time)/ current
(d) Work done Charge time

Answer
(a)
(**Hint:** By definition, potential difference or voltage = (work done / charge) and charge = current × time.)

11. How much energy is given to each coulomb of electric charge passing through a 6 V battery?

Answer
Charge $Q = 1$ C
Potential difference $V = 6$ V
Work done to move the charge through the battery, $W = VQ = 6$ V $\times$ 1 C = 6 J.
An amount of energy equal to the work done must be given to the charge.
So, the energy given to each coulomb of charge = 6 J.

12. Identify the circuit in which the electrical components have been properly connected.

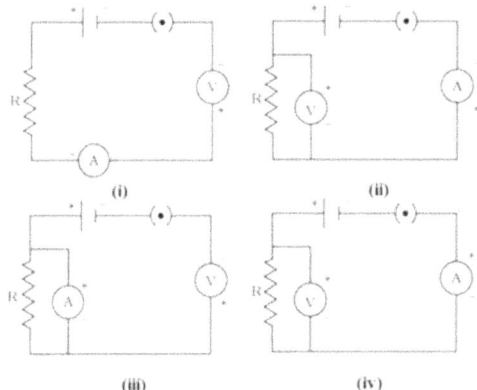

(a) (i)
(b) (ii)
(c) (iii)
(d) (iv)

Answer
(b)
(**Hint:** A voltmeter V must be joined *across* the resistance. Also, positive (+) and negative (-) terminals of both voltmeter and ammeter should be joined to the positive terminal and the (-) to the negative terminal respectively of the cell or the battery.)

13. Let the resistance of an electric component remain constant while the potential difference across the two ends of the component decrease to half its former value. What change will occur in the current through it?

Answer
The current of the electric component will decrease to half its former value.
(**Hint:** Ohm's law tells us that the current flowing through any conductor of given resistance is directly proportional to the potential difference across its two ends.)

14. A student carries out an experiment and plots the V-I graph of three samples of nichrome wire with resistances R_1, R_2 and R_3 respectively. Which of the following is true?

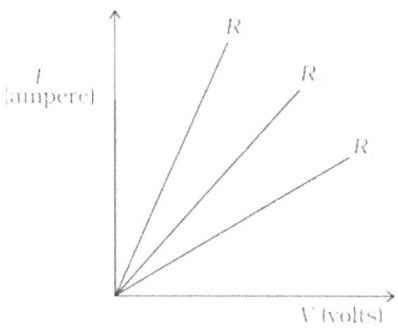

(a) $R_1 = R_2 = R_3$
(b) $R_1 > R_2 > R_3$
(c) $R_3 > R_2 > R_1$
(d) $R_2 > R_3 > R_1$

Answer
(c)
(**Hint:** Resistance is the ratio of potential difference to current (i.e., $R = V/I$). Note that since the current I is plotted along y-axis and V along x-axis, the (V/I) value of R_3 line is maximum and that of R_1 is minimum.)

15. State Ohm's law? How can it be verified experimentally? Does it hold good under all conditions?

Answer
Ohm's law states that the current flowing through a conductor is directly proportional to the potential difference across its two ends provided the temperature remains constant.
Complete the circuit as shown in the diagram given below. Here R denotes the nichrome wire.

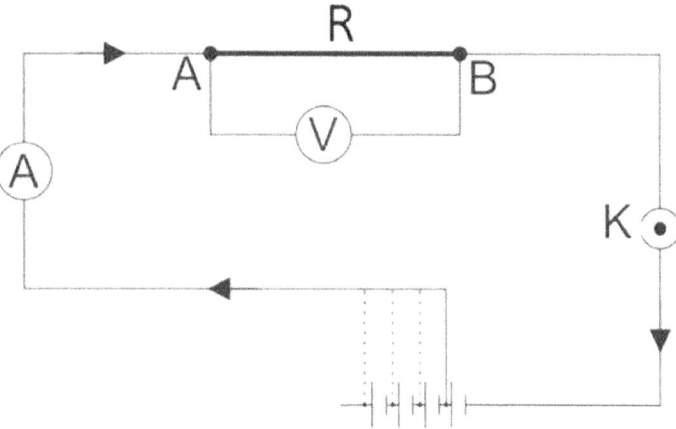

First, use only one cell in the circuit. Insert the plug key

Note the readings of the ammeter and the voltmeter V. Remove the key and connect two cells in series with the circuit. Reinsert the key and note the readings of the current I in the ammeter and the potential difference in the voltmeter. List the new values in the table.

Repeat the experiment by bringing in three cells and finally, all the four cells in the circuit, measuring and tabulating the values or I and V each time.

Each time calculate the ratio of V to I.

Plot the graph of the potential difference V against the current valves I. The ratio (V/I) is found to be almost equal in various steps of the experiment. The V-I graph is found to be a straight line passing through the origin O. These results are consistent with Ohm's law.

Ohm's law holds good only in case of metallic conductors whose temperature remains constant.

16. An electric circuit consisting of a 1.0 m long metallic wire XY, an ammeter, a voltmeter, 4 cells of 1.5 V each and a plug key was set up.
Draw a schematic diagram of this electric circuit in the 'on' position.
Following graph was plotted between the values of potential difference (V) and current (I). What conclusion do you draw about the relation between V and I from this graph. State this relation in your words.

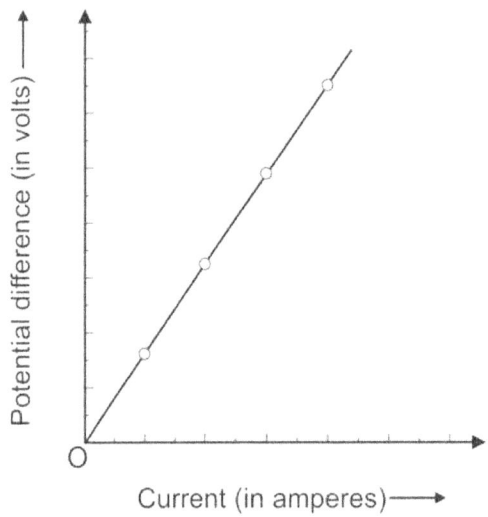

Answer
Schematic diagram of this electric circuit is similar to that of shown in Figure 4.1.
The V-I graph is a straight line passing through the origin O.
This result shows that the ratio V/I = constant or, I is proportional to V.
We can say that the current flowing through a conductor is directly proportional to the potential difference across its two ends.

17. The values of current I flowing in a given resistor for the corresponding values of potential difference V across the resistor are given by
I (A) : 0.5, 1.0, 2.0, 3.0, 4.0
V (V): 1.6, 3.4, 6.7, 10.2, 13.2
Plot the graph between V and I and calculate the resistance of the resistor.

Answer
Plot V along the y-axis (scale: 1cm = 2V). Plot I along x-axis (scale: 1cm = 1A). The graph as given below is a straight line.

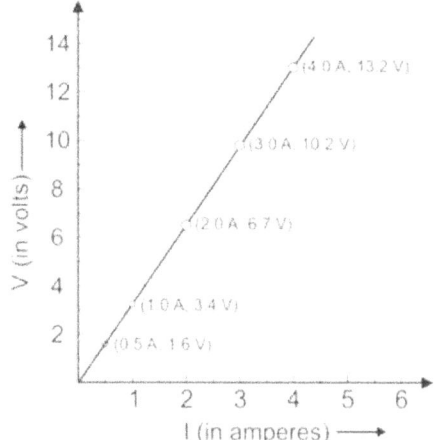

The resistance R is the ratio (V/I). From the graph, we find that for $V = 13.2$ V, $I = 4.0$ A. So, $R = 13.2$ V/ 4.0 A $= 3.3$ Ω.

18. A cell, a resistor, a key and ammeter are arranged as shown in the circuit diagrams given below. The current recorded in the ammeter will be

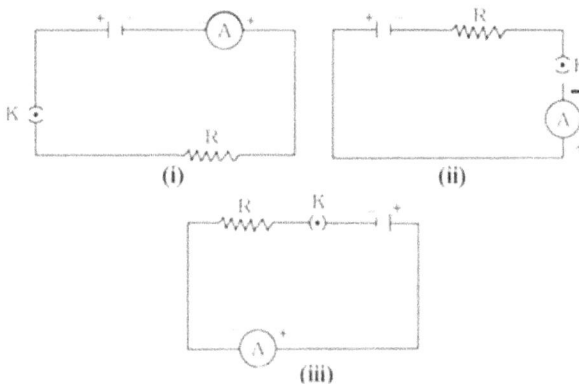

(a) maximum in (i)
(b) maximum in (ii)
(c) maximum in (iii)
(d) the same in all the cases

Answer
(d)
(**Hint:** In these diagrams, the only difference is the order of components and direction of current. But in all three cases, a resistance R is in series with the same cell and the same ammeter. So, the current must be same in all cases.)

19. Should the resistance of an ammeter be low or high? Give reason.

Answer
It should be as close to zero as possible. An ammeter is connected in series with the circuit through which the current to be measured flows. If the resistance of the ammeter is high, it will add on to the resistance of the circuit. Consequently, the true value of current will change.

20. When a 12 V battery is connected across an unknown resistor, there is a current of 2.5 mA in the circuit. Find the value of resistance of the resistor.

Answer
$V = 12$ V
$I = 2.5$ mA $= 2.5 \times 10^{-3}$ A
Ohm's law gives
Resistance, $R = V/I = 12$ V$/ (2.5 \times 10^{-3}$ A$) = 4.8 \times 10^{3}$ Ω.

21. What will be the value of current through the resistor, when a 24 V battery is connected across a 1000 Ω resistor?

Answer
$V = 24$ V
$R = 1000$ Ω
Ohm's law gives
Current $I = V/R = 24$ V$/ (1000$ $\Omega) = 2.4 \times 10^{-2}$ Ω.

22. Draw a schematic diagram of the circuit consisting of a battery of three cells of 2V each; a 5 Ω resistor, an 8 Ω resistor, a 12 Ω resistor, and a plug key, all connected in series.

Answer
The required circuit diagram is given below.

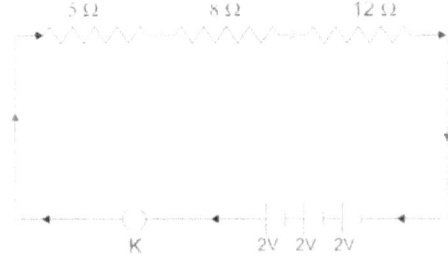

23. Redraw the circuit of Question 1, putting in an ammeter to measure the current through the resistors and a voltmeter to measure the potential difference across the 12 Ω resistor. What would be readings in the ammeter and the voltmeter?

Answer
The required circuit diagram is given below.

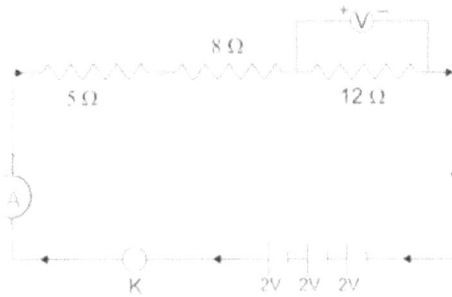

Total potential difference $V = 3 \times (2$ V$) = 6$ V.
The total resistance R is
$R = 5$ $\Omega + 8$ $\Omega + 12$ $\Omega = 25$ Ω.
So, ammeter reading $I = V/R = 6$ V$/25$ $\Omega = 0.24$ A.
Same current I flows through the 12 Ω resistor.
So, voltmeter reading $= I \times (12$ $\Omega) = (0.24$ A$) (12$ $\Omega) = 2.88$ V.

24. A child has drawn the electric circuit to study Ohm's law as shown in the diagram given below. His teacher told that the circuit diagram needs correction. Study the circuit diagram and redraw it after making all corrections.

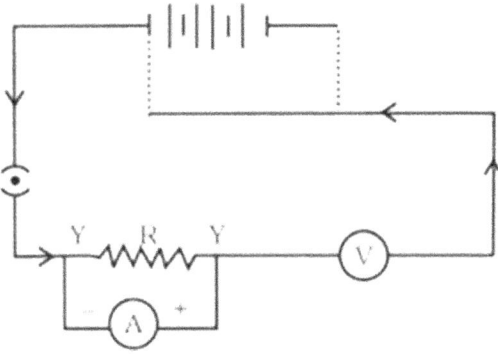

Answer
In the electric circuit drawn by the child, the following mistakes are present: (i) voltmeter is connected in series (ii) the cells are not joined in series (iii) ammeter is joined in parallel (iv) the current is shown flowing in the wrong direction.
In the correct circuit diagram given below, voltmeter is joined in parallel across the resistance and ammeter is joined in series. The cells are in series and the direction of current is correct.

25. (a) Name an instrument that measures electric current in a circuit. Define the unit if electric current.
(b) What do the following symbols mean in circuit diagrams?

(c) An electric circuit consisting of a 0.5 m long nichrome wire XY, an ammeter, a voltmeter, four cells of 1.5 V each and a plug key was set up.
(i) Draw a diagram of the electric circuit to study the relation between the potential difference maintained between the points 'X' and 'Y' and the electric current flowing through XY.
(ii) Following graph was plotted between V and I values

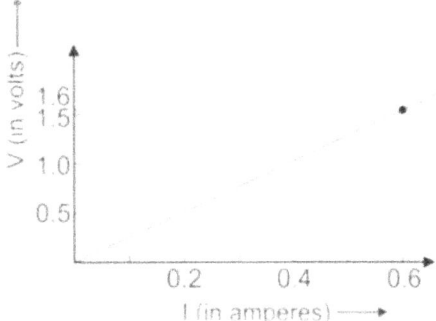

What would be the value of V/I ratios when the potential difference is 0.8 V, 1.2 V and 1.6 V respectively?
What conclusion do you draw from these values?

Answer
(a) The instrument is ammeter. The SI unit of current (known as ampere) is that much current which flows through a cross-section of a conductor when one coulomb of charge flows through it in one second.
(b) (i) A variable resistor (ii) Closed plug key or switch.
(c) (i) The diagram of the electric circuit is shown in Figure 4.1 (Chapter 4).

(ii) For $V = 1.6$ V, $I = 0.6$ A. So, the ratio $V/I = 1.6$ V/ 0.6 A $= 2.67$ Ω. Since V-I graph is a straight line, V/I has same value ($= 2.67$ Ω) when V is 1.2 V and 1.6 V.
V/I = constant means that I is proportional to V. We conclude that the current flowing through a conductor is directly proportional to the potential difference across its ends.

26. On what factors does the resistance of a conductor depend?

Answer
Resistance of a conductor at a given temperature depends on
(i) its length
(ii) its area of cross-section and
(iii) nature of its material.

27. (a) Distinguish between the terms, electrical resistance and resistivity of a conductor.
(b) A copper wire of resistivity 1.6×10^{-8} ohm meter has a cross sectional area of 20×10^{-4} cm². Calculate the length of this wire required to make a 10 ohm coil.

Answer
(a) Electrical resistance of a conductor is the ratio of the potential difference across the conductor to the current flowing through it. Its value depends on the physical dimensions of the conductor.
Electrical resistivity is a property of the material of the conductor. It is the resistance per unit length per unit area of cross-section of that material. Its value is independent of the physical dimensions of the conductor.
(b) $R = 10$ Ω, $A = 20 \times 10^{-4}$ cm² $= 20 \times 10^{-8}$ m²,
$\rho = 1.6 \times 10^{-8}$ Ω m
$l = ?$
$R = \rho l / A$ gives
$l = R A / \rho$,
$= 10$ $\Omega \times (20 \times 10^{-8}$ m²$) / (1.6 \times 10^{-8}$ Ω m$)$
$= 125$ m.

28. A cylindrical conductor of length l and uniform area of cross-section A has resistance R. Another conductor of length $2l$ and resistance R of the same material has area of cross section
(a) $A/2$
(b) $3A/2$
(c) $2A$
(d) $3A$

Answer
(c)
(**Hint:** If resistivity of the material of the conductor is ρ, $R = \rho l_1 / A_1 = \rho l_2 / A_2$. Or $l_1 / A_1 = l_2 / A_2$ or $A_2 / A_1 = l_2 / l_1 = (2l/l) = 2$ or $A_2 = 2 A_1$.)

29. What is electrical resistivity? In a series electrical circuit comprising a resistor made up of a metallic wire, the ammeter reads 5 A. The reading of the ammeter decreases to half when the length of the wire is doubled. Why?

Answer
Electrical resistivity is a property of the material of the conductor. Its value is independent of physical dimensions of the conductor but varies with temperature. It is numerically equal to the resistance per unit length of a uniform conductor of that material that has a unit area of cross-section.
We have resistance $R = \rho l / A$. If length l is doubled keeping area of cross-section A unchanged, R is also doubled.
According to Ohm's law, current $I = V/R$
When R is doubled while V remains unchanged, current becomes $I/2$.

30. The resistivity does not change if

(a) the material is changed
(b) the temperature is changed
(c) the shape of the resistor is changed
(d) both material and temperature are changed

Answer
(c)
(**Hint:** Resistivity is independent of physical dimensions of the resistor but depends on the material and temperature of the resistor.)

31. What happens to resistance of a conductor when its area of cross-section is increased?

Answer
The resistance of a conductor, being inversely proportional to its area of cross-section, decreases when its area of cross-section is increased

32. *(Based on Higher Order Thinking Skills (HOTS))* **Electrical resistivity of some substances at 200^0 C are given below:**
Silver : 1.60×10^{-8} Ω m
Copper : 1.62×10^{-8} Ω m
Tungsten : 5.20×10^{-8} Ω m
Iron : 10.0×10^{-8} Ω m
Mercury : 94.0×10^{-8} Ω m
Nichrome : 100×10^{-6} Ω m
Answer the following questions in relation to them:
(i) Between silver and copper, which one is a better conductor? Why?
(ii) Which material would you advice to be used in electrical heating devices? Why?

Answer
(i) Silver is a better conductor than copper because its resistivity is lower than that of copper.
(ii) Nichrome should be used in electrical heating devices because value of its resistivity shows that it is an alloy and alloys do not oxidize or burn easily.

33. Why are the coils of electric toasters made of an alloy rather than a pure metal?
Answer
(i) Alloys do not oxidise or burn easily.
(ii) The resistance of wires made of alloys changes very little with temperature.
(iii) Alloys have higher values of resistivity than metals. So, a small length of wire made of an alloy is sufficient to have required value of resistance.
(**Hint:** This result in (iii) follows from the relation $l = R A /\rho$, or l is proportional to $1/\rho$, for given R and A.)

34. What is electrical resistivity of a material? What is its unit? Describe an experiment to study the factors on which the resistance of conducting wire depends.

Answer
Resistivity is numerically equal to the resistance of a wire of unit length having an unit area of cross-section. Its unit is ohm metre (Ω m). Its value is independent of physical dimensions of the conductor but depends on the material of the conductor and the temperature
Take four conductors in the shape of uniform wires: (i) wire W_1, made of nichrome having certain length l and area of cross-section A, (ii) wire W_2, made of nichrome having same thickness as W_1 but twice the length of W_1, (iii) wire W_3, again made of nichrome having same length as W_1 but thicker than W_1 (iv) wire W_4 made of copper but with same length and cross-section as W_1.

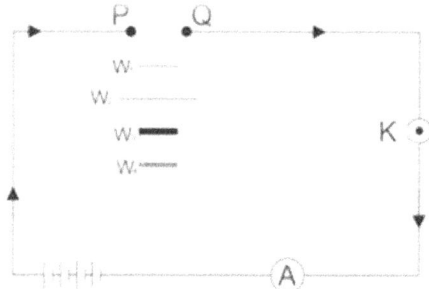

Complete the circuit as shown in the diagram given above.
Connect the wire W_1 in the gap PQ, so that the circuit is closed.
Insert the plug in the key. Note the reading of the ammeter. Let the current reading be I_1.
Next, replace W_1 with W_2 and again note the current I_2 in the ammeter.
In the same manner, replace W_2 with W_3 and note the value I_3 of the current.
We observe the value of current I_2 is half that of I_1.
The value I_3 is larger than that of I_1.
Finally, I_4 is different from I_1, which means that the resistance of W_4 (made of copper) is different from that of W_1.
We conclude that resistance of a conductor increases with increase of its length but decreases with increase of its cross-section. Moreover, it depends on the nature of the material of the conductor.

35. *(Numerical Problem Based on Higher Order Thinking Skills (HOTS))* **A piece of wire of resistance 20 Ω is drawn out so that its length is increased to twice its original length. Calculate the resistance of the wire in the new situation.**

Answer
Here, $(l_2 / l_1) = 2$.
Since volume (= length × area of cross-section) remains the same, $l_1 A_1 = l_2 A_2$, or, $(A_1 / A_2) = (l_2 / l_1) = 2$.
Now, $R_2 / R_1 = (\rho\, l_2 / A_2) / (\rho\, l_1 / A_1) = (l_2 / l_1)(A_1 / A_2) = (2) \times (2) = 4$.
So, $R_2 = 4 \times R_1 = 4 \times (20\,\Omega) = 80\,\Omega$.

36. A copper wire has a diameter of 0.5 mm and resistivity of 1.6×10^{-8} Ω m. What will be the length of this wire to make its resistance 10 Ω? How much does the resistance change if the diameter is doubled?

Answer
$d = 0.5$ mm $= 5 \times 10^{-4}$ m.
$R = 10\,\Omega$
$\rho = 1.6 \times 10^{-8}\,\Omega$ m
Length of wire $l = ?$
$R = \rho l / A$ gives
$l = R A / \rho = R (\pi d^2 / 4) / \rho$
or, $l = (10\,\Omega) \times (3.14) \times (5 \times 10^{-4}\,\text{m})^2 / (4) \times (1.6 \times 10^{-8}\,\Omega\,\text{m})$
$l = 122.7$ m.
From the relation
$R = \rho l / A$ $R = \rho l / (\pi d^2 / 4)$, we find that other quantities remaining constant, R is inversely proportional to (d^2)
So, when the diameter is doubled, resistance of wire will be one fourth of its earlier value i.e., $R' = R/4 = 10\,\Omega/4 = 2.5\,\Omega$.

37. In an electrical circuit comprising a resistor made up of a metallic wire, the ammeter reads 5 A. The reading of the ammeter is doubled when the length of the wire is halved. Why?

Answer
We have resistance $R = \rho l / A$. If length l is halved keeping area of cross-section A unchanged in case of the wire, R is also halved.

According to Ohm's law, current $I = V/R$
When R is halved, while V remains unchanged, current becomes twice its earlier value.

38. Alloys are used in electric heating devices rather than pure metals. Give one reason.
Answer
Alloys do not oxidize or burn easily at high temperature.

39. *(Based on Higher Order Thinking Skills (HOTS))* The electric resistivity of a few materials is given below in ohm-meter. Which of these materials can be used for making element of a heating device?
A : 6.84×10^{-8}
B : 1.60×10^{-8}
C : 1.00×10^{-4}
D : 2.50×10^{12}
E : 4.40×10^{-5}
F : 2.30×10^{17}

Answer
C and E.
(**Hint:** The resistivity values of C and E tell us that they are alloys.)

40. Electrical resistivity of a given metallic wire depends upon
(a) its length
(b) its thickness
(c) its shape
(d) nature of the material

Answer
(d)
(**Hint:** Resistivity is independent of physical dimensions of the conductor but depends on the material of the conductor.)

41. Will current flow more easily through a thick wire or thin wire of the same material, when connected to the same source? Why?

Answer
Current flows more easily through a thick wire. This is because a thick wire has lower resistance than a thin wire since resistance of a conductor is inversely proportional its area of cross-section.

42. Why are the coils of electric toasters and electric heaters made of an alloy rather than a pure metal?

Answer
(i) The wires made of alloys of metals do not get oxidised or burn easily.
(ii) The resistivity of an alloy being higher than that of metals by a factor of about 100, a small length of wire made of an alloy is sufficient to have required value of resistance.
(iii) Even at high temperature, the resistivity of alloys does not vary as much with temperature as that of pure metals.
(**Hint:** This result in (ii) follows from the relation $l = R A /\rho$, or l is proportional to $1/\rho$, for given R and A.)

43. Use the data in Table T 5.1 (chapter 5) of this book to answer the following-
(a) Which among iron and mercury is a better conductor?
(b) Which material is the best conductor?

Answer
(a) Iron.
(b) Silver.
(**Hint:** A material with lower resistivity is a better conductor.)

44. Judge the equivalent resistance when the following are connected in parallel
(a) 1 Ω and 10^6 Ω
(b) 1 Ω, 10^3 Ω and 10^6 Ω.

Answer
(a) We have
$1/R_p = 1/R_1 + 1/R_2 = 1/1 + 1/10^6 = 1 + 10^{-6} \approx$ slightly more than 1 Ω
So, R_p is slightly less than 1 Ω.
(b) We have
$1/R_p = 1/R_1 + 1/R_2 + 1/R_3 = 1 + 10^{-3} + 10^{-6} \approx$ slightly more than 1 Ω
So, R_p is slightly less than 1 Ω.
(**Note:** The above results show that when two or more than two resistances, one among whom happens to be very small compared to the rest, are combined in parallel, the equivalent resistance is slightly less than the smallest resistances.)

45. *(Numerical Problem Based on Higher Order Thinking Skills (HOTS))* **Two lamps, one rated 60 W at 220 V and the other 40 W at 220 V, are connected in parallel to the electric supply at 220 V.**
(a) Calculate the current drawn from the electric supply.
(b) Calculate the total energy consumed by the two lamps together when they operate for one hour.

Answer
(a) Currents drawn by 60 W lamp and 40 W lamp are, respectively,
$I_1 = P_1/V = (60 \text{ W})/(220\text{V}) = (3/11)$ A,
$I_2 = P_2/V = (40 \text{ W})/(220\text{V}) = (2/11)$ A.
For a parallel combination, the total current $I = I_1 + I_2 = (5/11)$ A = 0.45 A
(c) The total power consumed by the lamps $P = VI = V \times (I_1 + I_2) = VI_1 + VI_2 = P_1 + P_2 = 100$ W. So, total energy consumed in one hour = 100 W × 1 h = 100 W h = 0.1 kWh.

46. In the circuit diagram given below: calculate
(a) the current through each resistor.
(b) the total current in the circuit
(c) the total effective resistance of the circuit.

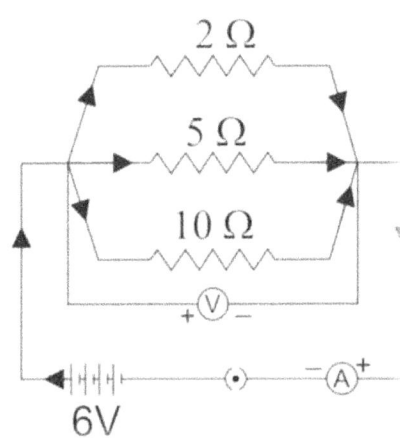

Answer
(a) The common potential difference across the parallel combination (V) = 6 V. So, the currents through 2 Ω, 5 Ω and 10 Ω resistors are respectively,
$I_1 = (6 \text{ V})/(2 \text{ Ω}) = 3.0$ A, $I_2 = (6 \text{ V})/(5 \text{ Ω}) = 1.2$ A, $I_3 = (6 \text{ V})/(10 \text{ Ω}) = 0.6$ A.
(b) The total current $I = I_1 + I_2 + I_3 = 3.0$ A + 1.2 A + 0.6 A = 4.8 A
(c) The total effective resistance is
$R_p = 6$ V/ (the total current) = 6V/ 4.8 A = 1.25 Ω.
(**Note**: R_p is also obtainable as $1/R_p = 1/R_1 + 1/R_2 + 1/R_3 = 1/2 + 1/5 + 1/10 = (8/10)$, or $R_p = 1.25$ Ω.)

47. How will you conclude that the same potential difference (voltage) exists across three resistors connected in a parallel arrangement to a battery?

Answer
Join the three resistances R_1, R_2 and R_3 in parallel and complete the circuit as shown in the diagram given below.

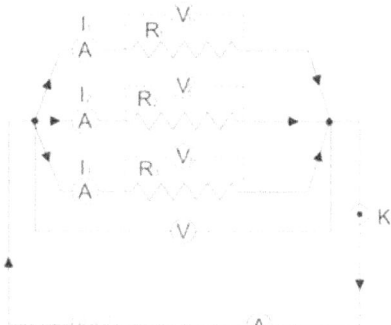

Insert the key and note the readings of the ammeter and that of the voltmeter. Let these values be I and V respectively Remove the key. Change the position of the voltmeter and connect it across R_1. Record the reading of the voltmeter. Let it be V_1.
Now, connect the voltmeter across R_2 and then across R_3. In each case, the key should be removed before changing the connection and reinserted before taking the reading. Let the corresponding readings of the voltmeter be V_2 and V_3 respectively.
We find that $V = V_1 = V_2 = V_3$, which shows that in a parallel combination, a common potential difference exists across each of the resistances.

48. (Numerical Problem Based on Higher Order Thinking Skills (HOTS)) Two students perform the experiments on series and parallel combination of two resistors R_1 and R_2 and plot the following V-I graphs Which of the graphs is (are) correctly labelled in terms of words 'series' and 'parallel'. Justify your answer.

Answer
We know that the equivalent resistance of a series combination of the two resistors is more than that of a parallel combination. In both cases, resistance is given by the ratio (V/I). This ratio (V/I) is more for a series combination than for a parallel combination in both the graphs. So, both graphs are correctly labelled. The justification is as follows.
Note that the first student plots V along y-axis and I along x-axis. So, the ratio (V/I) of series - line is more than that of the parallel-line. On the other hand, since the second student plots I along y-axis and V along x-axis, the ratio (V/I) of the parallel –line is less than that of the series –line.

49. (Numerical Problem Based on Higher Order Thinking Skills (HOTS)) A piece of wire having resistance R is cut into four equal parts.
(a) How does the resistance of each part compare with the original resistance?
(b) If the four parts are placed in parallel, how will the resistance of combination compare with the resistance of original wire?

Answer
(a) Since resistance of the wire is directly proportional to its length, resistance of one part = $R/4$.
(b) For parallel combination, $1/R' = 1/R/4 + 1/R/4 + ...$ 4 times $= 16/R$
$R' = R/16$. So, resistance of combination becomes $(1/16)$ times that of original wire.

50. For the circuit shown in the diagram given below, calculate:

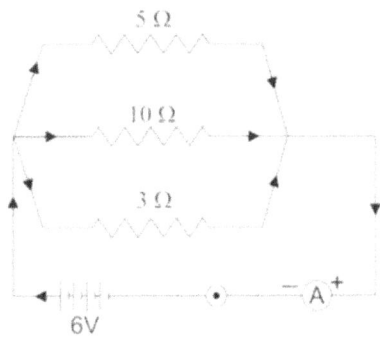

(a) the value of current through each resistor
(b) the total current in the circuit
(c) the total effective resistance of the circuit.

Answer
(a) Let $V (= 6$ V$)$ be the common potential difference across the parallel combination. Then, the currents through 5 Ω, 10 Ω and 30 Ω resistors are respectively,
$I_1 = V/R_1 = (6$ V$)/(5$ Ω$) = 1.2$ A,
$I_2 = V/R_2 = 0.6$ A,
$I_3 = V/R_3 = 0.2$ A
(b) For a parallel combination, the total current $I = I_1 + I_2 + I_3 = 2.0$ A
(c) The total effective resistance or the equivalent resistance is given by
$1/R_p = 1/R_1 + 1/R_2 + 1/R_3 = 1/5 + 1/10 + 1/30 = (10/30)$.
Or, $R_p = 3.0$ Ω.
(**Note**: I is also $= V/R_p = (6$ V$)/(3$ Ω$) = 2.0$ A, as expected for a parallel combination.)

51. Series arrangements are not used for domestic circuits. List any three reasons.

Answer
(i) Since a common current flows through the whole circuit, electric devices that need different values of current to operate cannot be used simultaneously in a series circuit,.
(ii) In case of series circuit, there is a common path through which the current flows. Consequently in case of failure of one of the defective devices, none of the other devices will work even though they may be in good working condition.
(iii) The equivalent resistance of a series circuit being high, the total current from a given power supply has a lower value. The low value of current may be too small for the proper working of some of the devices.

52. The proper representation of series combination of cells obtaining maximum potential is

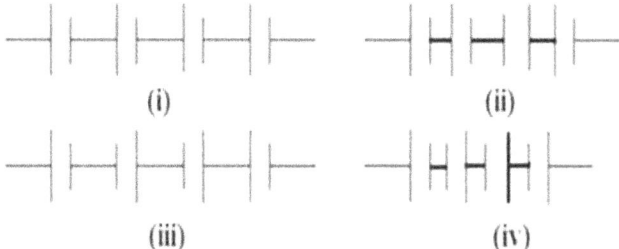

(a) (i)
(b) (ii)
(c) (iii)
(d) (iv)

Answer
(a)
(**Hint:** For obtaining maximum potential, cells need to be joined in series. In a series combination of cells, the negative terminal of each cell should be joined to positive terminal of next cell.)

53. Why is a series arrangement not used for connecting domestic electrical appliances in a circuit?
Answer
In a series arrangement, we cannot simultaneously use electric devices having different current ratings. Also, when one of the devices fails due to some defect, all other devices joined in series with it stop working.

54. A hot plate of an electric oven connected to a 220 V line has two resistance coils A and B, each of 24 Ω resistance, which may be used separately, in series or in parallel. What are the currents in the three cases?

Answer
Resistance R of each coil = 24 Ω
When the coils are used separately,
Current through each is
$I = V/R = 220 \text{ V}/24 \text{ Ω} = 9.2$ A.
When the coils are used in series,
Equivalent resistance is
$R_s = R + R = 2R = 48$ Ω
Current through the series combination
$I_s = V/R_s = 220 \text{ V}/48 \text{ Ω} = 4.6$ A.
When the coils are used in parallel,
Equivalent resistance is
$R_p = R \times R / (R + R) = R/2 = 12$ Ω
Current through the parallel combination
$I_p = V/R_p = 220 \text{ V}/12 \text{ Ω} = 18.3$ A.

55. Two resistors of resistance 2 Ω and 4 Ω when connected to a battery will have
(a) same current flowing through them when connected in parallel
(b) same current flowing through them when connected in series
(c) same potential difference across them when connected in series
(d) different potential difference across them when connected in parallel

Answer
(b)
(**Hint:** When resistors are connected in series, the same current flows through all of them.)

98

56. *(Numerical Problem Based on Higher Order Thinking Skills (HOTS))* **Two metallic wires A and B of same metal are connected in parallel. Wire A has length *l* and radius *r*, wire B has length 2*l* and radius 2 *r*. Compute the ratio of the total resistance of parallel combination and the resistance of wire A.**

Answer
If ρ be the resistivity of the metal, $R_1 = \rho\, l / A = \rho\, l / (\pi r^2)$,
$R_2 = \rho\, (2l)/ (\pi (2r)^2) = R_1/2$.
Now, equivalent resistance of parallel combination $R_p = R_1 R_2 / (R_1 + R_2) = (R_1)(R_1/2)/(R_1 + R_1/2) = (1/3) \times R_1$.
So, $R_p/R_1 = 1/3$.

57. Find out the following in the electric circuit given below

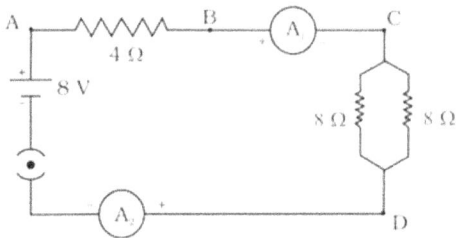

(a) **Effective resistance of two 8 Ω resistors in the combination**
(b) **Current flowing through 4 Ω resistor**
(c) **Potential difference across 4 Ω resistance**
(d) **Power dissipated in 4 Ω resistor**
(e) **Difference in ammeter readings, if any.**

Answer
(a) Effective resistance of the parallel combination is $R_p = R_1 \times R_2 / (R_1 + R_2) = (8\times 8)/(8+8)\ \Omega = 4\ \Omega$.
(b) The total resistance in the circuit is given by
$R_T = 4\ \Omega + R_p = 4\ \Omega + 4\ \Omega = 8\ \Omega$.
To get current flowing through 4 Ω resistor we can apply Ohm's law to get $I = V/R_T = (8\ V)/(8\ \Omega)$.
Or $I = 1.0$ A.
(c) Potential difference across 4 Ω resistance is
$V_1 = I R_1 = 1.0\ A \times 4\ \Omega = 4V$.
(d) Power P is given as
$P = I^2 R = (1.0\ A)^2 \times (4\ \Omega) = 4$ W.
(e) There is no difference in ammeter readings because same current flows through all components in series.

58. What is the minimum resistance which can be made using five resistors each of 1/5 Ω
(a) 1/5 Ω
(b) 1/25 Ω
(c) 1/10 Ω
(d) 25 Ω

Answer
(b)
(**Hint:** To obtain minimum resistance, we need to join all the resistances in parallel. Let R_p be the equivalent resistance, we have $1/R_p = 1/R + 1/R + ...$ 5 times $= 1/(1/5) + 1/(1/5) + ...$ 5 times $= 5 \times (5/1) = 25/1$ or $R_p = (1/25)\ \Omega$.)

59. *(Numerical Problem Based on Higher Order Thinking Skills (HOTS))* **Show how you would connect three resistors, each of resistance 6 Ω, so that the combination has a resistance of (i) 9 Ω (ii) 4 Ω.**

Answer
(i) The desired arrangement is shown in Figure (a) given below.

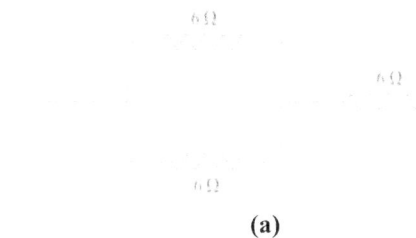

(a)

The equivalent resistance R' of parallel combination of 6 Ω and 6 Ω is
$R' = (6\,\Omega) \times (6\,\Omega) / (6\,\Omega + 6\,\Omega) = 3\,\Omega$.
So, the equivalent resistance of the whole combination (Figure (b)) is

$R\,(=3\,\Omega)$ $6\,\Omega$

(b)

$R = R' + 6\,\Omega = 3\,\Omega + 6\,\Omega = 9\,\Omega$, as desired.
(ii) After a little hit and trial, we find that the arrangement given in Figure (c) gives the desired result.

(c)

The equivalent resistance R' of series combination of 6 Ω and 6 Ω is
$R' = 6\,\Omega + 6\,\Omega = 12\,\Omega$.
So, the equivalent resistance of the whole arrangement (Figure (d)) is

$R\,(=12\,\Omega)$

$6\,\Omega$

(d)

$R = (6\,\Omega) \times (R') / (6\,\Omega + R') = (6\,\Omega) \times (12\,\Omega) / (6\,\Omega + 12\,\Omega) = 4\,\Omega$, as desired.
(**Hint:** (i) Note that in a series combination the equivalent resistance is larger the largest resistance. So, to get a resistance of 9 Ω, one of the three resistors of 6 Ω can be one of the resistances in the series. The remaining 3 Ω (= 9 Ω - 6 Ω) is obtainable by joining the other two given resistors in parallel.
(ii) To obtain a resistance of 4 Ω, note that the equivalent resistance being larger than any one of the resistances in a series combination, none of the three resistors of 6 Ω can be part of any series combination. So, one of them should be in parallel with a combination of the other two resistors. The new combination of two resistors can either be series or parallel. Rough calculation of the two possibilities tells us that these two resistances should be put in series.)

60. What is the maximum resistance which can be made using five resistors each of 1/5 Ω

(a) 1/5 Ω
(b) 10 Ω
(c) 5 Ω
(d) 1 Ω

Answer
(d)
(**Hint:** To obtain maximum resistance, we need to join the resistances in series. Let R_s be the equivalent resistance of the series combination. We have $R_s = 1/5\,\Omega + 1/5\,\Omega + ...$ 5 times $= (5/5)\,\Omega = 1\,\Omega$.)

61. How will you infer with the help of an experiment that the same current flows through every part of the circuit containing three resistances in series connected to a battery?

Answer
Join the three resistances R_1, R_2 and R_3 in parallel and complete the circuit as shown in the diagram given below.

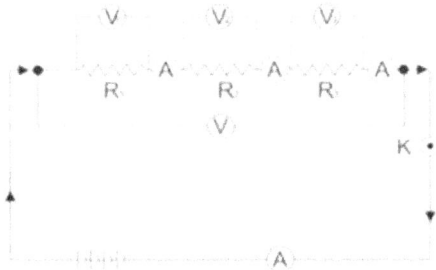

Insert the key and note the readings of the ammeter and that of the voltmeter. Let these values be I and V respectively. Remove the key. Change the position of the ammeter and connect it between R_1 and R_2. Record the reading of the ammeter. Let it be I_1.
Now, connect the ammeter between R_2 and R_3 and then between R_3 and the key. Each time, remove the key before changing the connection and reinsert it before taking the reading. Let the corresponding readings of the ammeter be I_2 and I_3 respectively. We find that $I = I_1 = I_2 = I_3$. It shows that the same current flows through every part of the circuit.

62. (Based on Higher Order Thinking Skills (HOTS)) A piece of wire of resistance R is cut into five equal parts. These parts are then connected in series. If the equivalent resistance of this combination is R', then the ratio R/R' is
(a) 1/25
(b) 1/5
(c) 5
(d) 1

Answer
(d)
(**Hint:** Resistance of one part $= R/5$. Now, $R' = R/5 + R/5 + ...$ 5 times $= 5 \times R/5 = R$. So, $R' = R$ and $R/R' = 1$.)

63. Explain the following:
(a) Why is the tungsten used almost exclusively for filament of electric lamps?
(b) Why are the conductors of electric heating devices, such as bread-toasters and electric irons, made of an alloy rather than a pure metal?
(c) Why is the series arrangements not used for domestic circuits?
(d) How does the resistance of a wire vary with its area of cross-section?
(e) Why are copper and aluminium wires usually employed for electricity transmission?

Answer

(a) Tungsten has very high melting point (3380° C) and high resistivity.
(b) The reasons are as follows.
(i) Alloys do not get oxidised or burn easily.
(ii) The resistance of wires made of alloys changes very little with temperature.
(iii) The resistivity of an alloy being higher than that of metals, we need a smaller length of wire made of an alloy than the one made of metal to have required value of resistance.
(c) Listed below are the main reasons
(i) Electric devices that need different values of current to operate cannot be used simultaneously in a series circuit, because a common current flows through the whole circuit.
(ii) In case of series circuit, there is a common path through which the current flows. Consequently in case of failure of one of the defective devices, none of the other devices will work even though they may be in good working condition.
(iii) The equivalent resistance of a series circuit being high, the total current from a given power supply has a lower value. The low value of current may be too small for the proper working of some of the devices.
(d) The resistance of a conductor varies inversely as its area of cross-section.
(e) Copper and aluminium are used for electricity transmission because they are very good conductors.

64. Which of the following terms does not represent electrical power in a circuit?
(a) I^2R
(b) IR^2
(c) VI
(d) V^2/R

Answer
(b)

65. *(Numerical Problem Based on Higher Order Thinking Skills (HOTS))* In the circuit diagram shown, the two wires X and Y are of the same length and same metal, but X is thicker than Y. Which ammeter A_1 or A_2 will indicate a higher current strength?

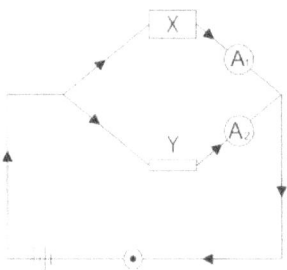

Answer
The wire X, being thicker, has lower resistance than that of Y. For a parallel combination, there is a common potential difference across X and Y. So, more current flows through X than through Y. Hence, A_1 will indicate higher value.

66. Three resistors of 10 Ω, 15 Ω and 5 Ω are connected in parallel. Find their equivalent resistance.

Answer
(i) $R_1 = 10$ Ω, $R_2 = 15$ Ω and $R_3 = 5$ Ω.
Let R_p be the equivalent resistance, we have
$1/R_p = 1/R_1 + 1/R_2 + 1/R_3$
$= 1/10 + 1/15 + 1/5 = 11/30$
or $R_p = (30/11)$ Ω $= 2.73$ Ω.

67. Draw a circuit diagram of an electric circuit containing a cell, a key, an ammeter, a resistor of 2 Ω in series with a combination of two resistors (4 Ω each) in parallel and a voltmeter across the parallel combination. Will the potential difference across the 2 Ω resistor be the same as that across the parallel combination of 4 Ω resistors? Give reason.

Answer
The circuit diagram is given below.

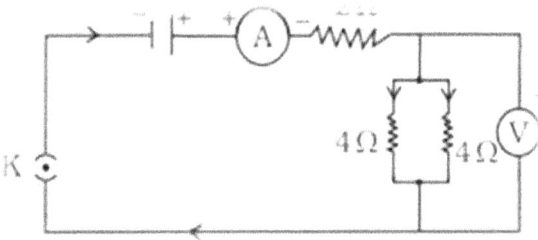

Yes. The reason is as follows. .
Let I be the total current in the circuit. Then the potential difference across the 2 Ω resistor = 2 I. Also, equivalent resistance of the parallel combination = (4×4)/ (4+4) Ω = 2 Ω. So, the potential difference across the parallel combination of 4 Ω resistors = 2 I.

68. A battery of 9 V is connected in series with resistors of 0.2 Ω, 0.3 Ω, 0.4 Ω, 0.5 Ω and 12.0 Ω respectively. How much current would flow through the 12 Ω resistor?

Answer
$V = 9$ V
Equivalent resistances is
$R_s = R_1 + R_2 + R_3 + R_4 + R_5 = 0.2$ Ω $+ 0.3$ Ω $+ 0.4$ Ω $+ 0.5$ Ω $+ 12.0$ Ω $= 13.4$ Ω.
Current I through the series combination
$I = V/R_s = 9$ V$/13.4$ Ω $= 0.67$ A.
In series, same current I flows through each of the resistances. .Hence, the current through 12 Ω resistance is 0.67 A.

69. *(Based on Higher Order Thinking Skills (HOTS))* Two conducting wires of the same material and of equal lengths and equal diameters are first connected in series and then parallel in a circuit across the same potential difference. The ratio of heat produced in series and parallel combinations would be
(a) 1:2
(b) 2:1
(c) 1:4
(d) 4:1

Answer
(c)
(**Hint:** Let the resistance of one conducting wire $=R$. In series $R_s = R + R = 2R$. In parallel, $R_p = R \times R/ (R + R) = R/2$. So, $H_s/ H_p = (V^2 t/ R_s) / (V^2 t / R_p) = R_p/ R_s = 1/4$.)

70. A current of 1 ampere flows in a series circuit containing an electric lamp and a conductor of 5 Ω when connected to a 10 V battery. Calculate the resistance of the electric lamp. Now if a resistance of 10 Ω is connected in parallel with this series combination, what change (if any) in current flowing through 5 Ω conductor and potential difference across the lamp will take place? Give reason.

Answer
Current in the series circuit = 1 A. Resistance of the conductor = 5 Ω.
Total resistance in the series circuit R_T = resistance of the conductor + resistance of the lamp = 10 V/1 A = 10 Ω.
So, the resistance of the electric lamp = R_T - 5 Ω = 10 Ω - 5 Ω = 5 Ω.
When a resistance of 10 Ω is connected in parallel with this series combination, effective resistance of the changed circuit R_p
= $R_1 \times R_2 / (R_1 + R_2) = (10 \times 10)/ (10+10)$ Ω $= 5$ Ω.
The changed current in the new circuit = $V/ R_p = 10$ V/5 Ω = 2A.

So, current passes through a parallel combination of R_T and additional resistance of 10 Ω = 2 A. Since R_T is also 10 Ω, the current of 2 A is divided equally between of R_T and the additional resistance of 10 Ω.
Thus the current through resistors R_T and additional resistance of 10 Ω each = (2/2) = 1 A. This current of 1 A passes through both electric lamp and the conductor of resistance 5 Ω. Hence there will be no change in current flowing through 5 Ω conductor. Since current through electric lamp continues to be 1 A, there will be no change in potential difference across the lamp either.

71. *(Numerical Problem Based on Higher Order Thinking Skills (HOTS))* **An electric lamp of resistance 100 Ω, a toaster of resistance 50 Ω and a water filter of resistance 500 Ω are connected in parallel to a 220 V source. What is the resistance of an electric iron connected to the same source that takes as much current as all three appliances and what is the current through it?**

Answer
The equivalent resistance of the parallel combination is given as
$1/R_p = 1/R_1 + 1/R_2 + 1/R_3$
= 1/100 + 1/50 + 1/500 = 4/125.
Or R_p = (125/4) Ω = 31.25 Ω.
The total current drawn by all three appliances is
$I_T = V/R_p$ = (220 V) / (31.25 Ω) = 7.04 A.
So, current through the electric iron = 7.04 A and resistance of the electric iron
= (220 V) / (7.04 A) = 31.25 Ω, which equals resistance of the parallel combination, as expected.

72. What are the advantages of connecting devices in parallel with the battery instead of connecting them in series?

Answer
(i) In a parallel combination, devices having different current ratings but same voltage rating can be connected in different branches of the circuit.
(ii) In a parallel combination, devices in good condition keep on working in other branches even if current stops flowing in one branch due to a defective device.
(iii) Due to lower equivalent resistance of a parallel combination, we have higher value of total current. So, it is possible for devices in different branches to have sufficient value of current for their proper functioning.

73. *(Numerical Problem Based on Higher Order Thinking Skills (HOTS))* **How can three resistors of resistance 2 Ω, 3 Ω and 6 Ω be connected to give a total resistance of (a) 4 Ω (b) 1 Ω?**

Answer
(a) To obtain a total resistance of 4 Ω, we should connect the two resistances of 3 Ω and 6 Ω in parallel and the join this parallel combination in series with resistance of 2 Ω. Such a combination is shown below.

To verify, first note that the equivalent resistance of parallel combination is
R' = (3 Ω) × (6 Ω) / (3 Ω + 6 Ω) = 2 Ω.
So, the equivalent resistance of the whole combination is
$R = R' + 2$ Ω = 2 Ω + 2 Ω = 4 Ω.
(b) To obtain the resistance of 1 Ω, join all the three resistances in parallel as shown below.

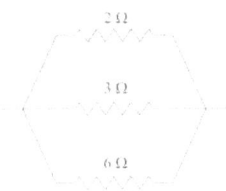

If R be the equivalent resistance of the parallel combination, we have
$1/R = 1/2 + 1/3 + 1/6 = (6/6)$,
, which gives $R = 1\ \Omega$, as desired.

(**Hint:** (i) note that in series combination the equivalent resistance is larger the largest resistance. So, to get a resistance of 4 Ω, the resistance of 6 Ω cannot be one of the resistances in the series combination. Moreover, in a parallel combination, the equivalent resistance is smaller than the smallest resistance of 2 Ω. So, all the resistances cannot be joined in parallel. Consequently, two of the resistances (one of which must be the resistance of 6 Ω) need to be joined in parallel.)
(ii) The equivalent resistance of a parallel combination is smaller than the smallest of all the resistances of the combination i.e., smaller than 2 Ω in the present case.)

74. What is (a) the highest (b) the lowest total resistance that can be secured by combining four coils of resistance values 4 Ω, 8 Ω, 12 Ω and 24 Ω?

Answer
(a) To obtain the highest value, we need to join the resistances in series. Then, the equivalent resistance is
$R_s = 4\ \Omega + 8\ \Omega + 12\ \Omega + 24\ \Omega = 48\ \Omega$. So, highest value= 48 Ω.
(b) To obtain the lowest value, we need to join all the resistances in parallel.
The equivalent resistance is given by
$1/R_p = 1/4 + 1/8 + 1/12 + 1/24 = 1/2$
or $R_p = 2\ \Omega$. So, lowest value= 2 Ω.
(**Hint:** (a) The equivalent resistance of a series combination is the sum of the values of the individual resistances. Obviously, such a sum gives the highest value.
(b) The equivalent resistance of a parallel combination is smaller than the smallest of the individual resistances. Obviously, such a result gives the lowest value among the possible combinations.)

75. Why is parallel arrangement used in domestic wiring?

Answer
(i) Devices working at voltage equal to supply voltage but having different current ratings can be connected in different branches of a parallel circuit.
(ii) In a parallel combination, devices in good condition in other branches keep on working even if current stops in one branch due to a defective device.
(iii) In a parallel combination, higher value of total current (due to lower equivalent resistance) makes it possible for devices in different branches to have sufficient value of current for their proper functioning.

76. (Numerical Problem Based on Higher Order Thinking Skills (HOTS)) (a) The electric power consumed by a device may be calculated by using either of the two expressions $P = I^2 R$ or $P = V^2/R$. The first expression indicates that P is directly proportional to R where as the second expression indicates inverse proportionality. How can the seemingly difference dependence of P on R in these expression be explained?
(b) Explain the following:
(i) Why is tungsten used almost exclusively for filament of electric limps?
(ii) Why are copper and aluminium wires used for electricity transmission?

Answer

(a) $P = I^2 R$ tells us that when current I through different resistances is common, power P is directly proportional to the resistance. Thus, in a series circuit, the larger resistance consumes more power than a smaller resistance.
$P = V^2/R$ tells us that when V across different resistances is common, power P is inversely proportional to the resistance. Thus, in a parallel combination, the larger resistance consumes less power than a smaller resistance.
(b) (i) Tungsten has very high melting point and high resistivity.
(ii) Copper and aluminium are very good conductors.

77. (Numerical Problem Based on Higher Order Thinking Skills (HOTS)) Two resistors, with resistances 5 Ω and 10 Ω respectively are to be connected to a battery of emf 6 V so as to obtain : (i) minimum current flowing (ii) maximum current flowing
(a) How will you connect the resistance in each case?
(b) Calculate the strength of the total current in the circuit in the two cases.

Answer
(i) (a) For minimum current, the resistance should be joined in series. The equivalent resistance is $R_s = 5\,\Omega + 10\,\Omega = 15\,\Omega$.
(b) Minimum current
$= V/R_s = (6\text{ V})/(15\,\Omega) = 0.4$ A.
(ii) (a) For maximum current, the resistance should be joined in parallel.
The equivalent resistance is
$R_p = (5\,\Omega) \times (10\,\Omega)/(5\,\Omega + 10\,\Omega) = 3.33\,\Omega$
(b) Maximum current
$= V/R_p = (6\text{ V})/(3.33\,\Omega) = 1.8$ A.

78. An electric lamp of 200 Ω and a toaster of 100 Ω are connected in parallel to a 220 V electricity source.
(a) What will be the resistance of an electric iron which when connected to the same electric source permits the same current as the total current flowing through both the appliances described above?
(b) What is the current passing through the electric iron?
(c) Calculate the power of the electric iron.

Answer
(a) The equivalent resistance of the parallel combination is given by
$1/R_p = 1/R_1 + 1/R_2 = 1/200 + 1/100 = 3/200$.
Or $R_p = 66.7\,\Omega$.
The electric iron joined to the same source takes as much current as is taken by both the appliances. The resistance of the electric iron will, therefore, be equal to the equivalent resistance of the parallel combination i.e. 66.7 Ω.
(b) The total current drawn by the combination, $I_T = V/R_p = (220\text{ V})/(66.7\,\Omega) = 3.3$ A. The current through the electric iron will be the same as through the combination i.e. 3.3 A.
(c) Power of the electric iron $= V \times I_T = (220\text{ V}) \times (3.3\text{ A}) = 726$ W.
(Hint: A different method is as follows.
(a) and (b) Currents drawn by 200 Ω lamp and 100 Ω toaster are, respectively,
$I_1 = V/R_1 = (220\text{ V})/(200\,\Omega) = 1.1$ A, $I_2 = V/R_2 = (220\text{ V})/(100\,\Omega) = 2.2$ A.
Total current $I = I_1 + I_2 = 3.3$ A. The current through the electric iron will be the same as through the combination i.e. 3.3 A.
So, resistance of the electric iron connected to the same electric source $= V/I = (220\text{ V})/(3.3\text{ A}) = 66.7\,\Omega$.
(c) Power of the electric iron $= V \times I = (220\text{ V}) \times (3.3\text{ A}) = 726$ W.**)**

79. Compute the heat generated while transferring 96000 coulombs of charge in 1hr through a potential difference of 50 V.

Answer
Charge $Q = 96000$ C,
Potential difference $V = 50$ V
By definition, the amount of work done in moving a charge Q through a potential difference V is given by
$W = V \times Q$

In order to this work, the source must be supplying an equal amount of electrical energy to the circuit. Assuming that the circuit contains resistors only, the whole of the electric energy supplied to the circuit gets converted into the heat energy. Therefore, the heat H produced when charge Q flows through a potential difference V is
$H = W = V \times Q$
$= (96{,}000 \text{ C}) \times (50 \text{ V})$
$= 4.8 \times 10^6 \text{ J}.$

80. (Based on Higher Order Thinking Skills (HOTS)) Two conducting wires of the same material and of equal lengths and equal diameters are first connected in series and then parallel in a circuit across the same potential difference. The ratio of power consumed in series and parallel combinations would be
(a) 1:2
(b) 2:1
(c) 1:4
(d) 4:1

Answer
(c)
(**Hint:** Let the resistance of one conducting wire $=R$. In series $R_s = R + R = 2R$. In parallel, $R_p = R \times R/(R+R) = R/2$. Then, the formula for power $P = V^2/R$ tells us that for common V, power consumed in series is $P_s = V^2/R_s$, and in parallel, it is $P_p = V^2/R_p$. So, $P_s/P_p = (V^2/R_s)/(V^2/R_p) = R_p/R_s = (R/2)/(2R) = \frac{1}{4}.$)

81. Which uses more energy, a 250 W TV set in 1 hr or a 1200 W toaster in 10 minutes?

Answer
Electric energy = electric power × time
So the electric energy used by TV set
= 250 W × 1 h = 250 W h
Electric energy used by toaster
= 1200 W × 10 min = 1200 W × (1/6) h = 200 W h
Therefore TV set uses more energy.

82. An electric iron of resistance 20 Ω takes a current of 5 A. Calculate the heat develop in 30 s.

Answer
Here, $R = 20$ Ω,
$I = 5$ A and
$t = 30$ s.
So, heat developed, $H = I^2 R t$
$= (5 \text{ A})^2 \times (20 \text{ Ω}) \times (30 \text{ s}) = 15{,}000 \text{ J}.$

83. (Based on Higher Order Thinking Skills (HOTS)) Why does the cord of an electric heater not glow while the heating element does?

Answer
The heating element, being made of thin wire of an alloy has much higher value of resistance than the cord of the heater which is made of thick copper wire. According to Joule's law, when same current flows for the same time interval, the heat produced is directly proportional to the resistance. So, the heating element becomes much hotter than the cord and starts glowing whereas the cord does not.

84. Derive the expression for the heat produced due to a current 'I' flowing for a time interval 't' through a resistor 'R' having a potential difference 'V' across its ends. With which name is the relation known as? How much heat will an instrument of 12 W produce in one minute if it is connected to a battery of 12 V?

Answer

Work done in moving a charge Q is
$W = VQ = VIt$ (since, $Q = It$)
To do this work, the source supplies an equal amount of electrical energy, which gets converted into the heat energy.
So, heat H produced
$H = VIt$
Ohm's law gives
$V = RI$
Therefore,
$H = I^2 R t$
This relation is known as 'Joule's law' of heating.
Heat produced = electric energy = $P t$ = 12 W × 1 min = 12 W × 60 s = 720 W s = 720 J.

85. What is Joule's heating effect? List its four applications in daily life.

Answer
Joule's heating effect states that if a steady current I flows through a resistor R, then the amount of heat H produced in time t is given as
$H = I^2 R t$.
Applications of Joule's heating effect in daily life are: (i) electric heater (ii) electric iron, (iii) electric bulb (iv) electric fuse.

86. In the following circuits, heat produced in the resistor or combination of resistors connected to a 12 V battery will be

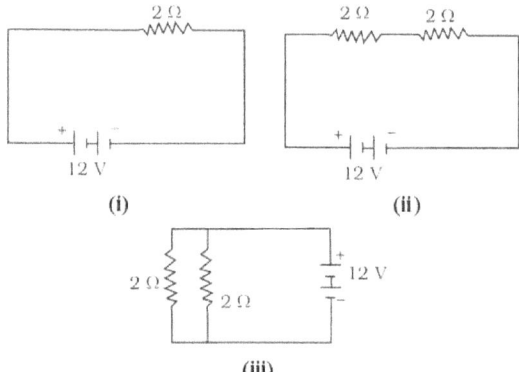

(a) same in all the cases
(b) minimum in case (i)
(c) maximum in case (ii)
(d) maximum in case (iii)

Answer
(d)
(**Hint:** In case (i), total resistance = 2 Ω, in case (ii), total resistance = (2+2) Ω = 4 Ω, in case (iii), total resistance = (2×2)/(2+2) Ω = 1 Ω. Now, heat produced per second or power $P = V^2/R$. Since R is minimum in case (iii), heat produced will be maximum in this case.)

87. How does use of a fuse wire protect electrical appliances?

Answer
The fuse is placed in series with the appliance to be protected so that the same current passes through it as through the appliance. When this current exceeds a certain safe value, due to Joule heating, the temperature of the fuse wire increases to its melting point. The fuse wire melts and the circuit breaks.

88. In an electrical circuit two resistors of 2 Ω and 4 Ω respectively are connected in series to a 6 V battery. The heat dissipated by the 4 Ω resistor in 5 s will be
(a) 5 J
(b) 10 J
(c) 20 J
(d) 30 J

Answer
(c)
(Hint: We know that in series combination, a common current I flows through each of the bulbs. We have $I = V/(R_1 + R_2) = 6V/(2+4) \, \Omega = 1.0$ A. So, heat dissipated by 4 Ω resistor = $I^2 R t = (1.0 \text{ A})^2 \times 4 \, \Omega \times 5 \text{ s} = 20$ J.)

89. *(Numerical Problem Based on Higher Order Thinking Skills (HOTS))* **A household uses the following electric appliances:**
(i) Refrigerator of rating 400 W for ten hours each day.
(ii) Two electric fans of rating 80 W for twelve hours each day.
(iii) Six electric tubes of rating 18 W for 6 hours each day.
Calculate the electricity bill of the household for the month of June if the cost per unit of electric energy is Rs. 3.00.

Answer
The electric energy used per day by
(i) refrigerator = 400 W × 10 h = 4000 W h
(ii) two electric fans = 2 × 80 W × 12 h = 1920 W h
(iii) six electric tubes = 6 × 18 W × 6 h = 648 W h.
So, the total energy consumed per day
= 4000 W h + 1920 W h + 648 W h = 6568 W h = 6.568 kW h = 6.568 units.
The total energy consumed for the month of June = 30 days × 6.568 units/day = 197.0 units.
Hence, the cost of electric energy = 197.04 units × Rs 3.0/ unit = Rs. 591

90. **Out of 60 W and 40 W lamps, which one has a higher electrical resistance when in use?**

Answer
Resistance R of the lamp is related to the power P as $R = V^2/P$. Or, R is inversely proportional to P. So, the 40 W lamp has higher resistance.

91. **B_1, B_2 and B_3 are three identical bulbs connected as shown in the diagram given below. When all the three bulbs glow, a current of 3A is recorded by the ammeter A.**

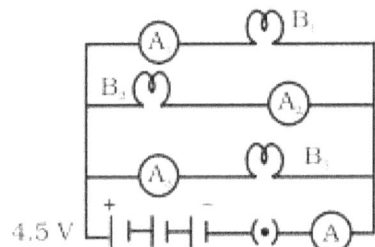

(i) **What happens to the glow of the other two bulbs when the bulb B_1 gets fused?**
(ii) **What happens to the reading of A_1, A_2, A_3 and A when the bulb B_2 gets fused?**
(iii) **How much power is dissipated in the circuit when all the three bulbs glow together?**

Answer

(i) B_1, B_2 and B_3 are joined in parallel combination. The glow of the bulbs B_2 and B_3 will remain the same when the bulb B_1 gets fused because in a parallel combination, current keeps on flowing in other branches even if current stops in one branch due to a defective device.

(ii) Since B_1, B_2 and B_3 have the same resistance, the total current of 3A is divided among the three branches containing B_1, B_2 and B_3 when all of them glow. Suppose R is the resistance of each of the bulbs. The equivalent resistance of this parallel combination R_p = 4.5 V/ 3A = 1.5 Ω.
Now, $1/R_p$= 1/R+ 1/R +1/R = 3/R or R_p = 1.5 Ω = R/3. So, R = 3 × 1.5 Ω = 4.5 Ω.
When the bulb B_2 gets fused, $1/R_p$= 1/R+ 1/R = 2/R. So, R_p = 4.5 Ω /2 = 2.25 Ω. So total current through A = 4.5V/ 2.25 Ω = 2 A. This total current of 2 A is divided equally between the branches containing B_1 and B_3. So both A and A_3 show 1 ampere (=2 A/2) each, whereas A_2 shows zero and A shows 2 ampere.

(iii) When all the three bulbs glow together, total current of 3A flows through the circuit.
Power $P = VI$ = 4.5 V × 3 A = 13.5 W

92. An electric room heater is rated at 2kW. Calculate the cost of using it for 2 hours daily for the month of September, if each unit costs Rs. 4.00.

Answer
Electric energy used per day = 2 kW × 2 h = 4 kW h. = 4 units.
The total energy consumed for the month of September = 30 days × 4 units/day = 120 units.
Hence, the cost of electric energy = 120 units × Rs. 4.0/ unit = Rs. 480.

93. (Numerical Problem Based on Higher Order Thinking Skills (HOTS)) Several electric bulbs designed to be used on a 220 V electric supply line, are rated 10 W. How many lamps can be connected in parallel with each other across the two wires of 220 V line if the maximum allowable current is 5 A?

Answer
Number of branches = the required number of bulbs = n (say). Using $P = VI$ or $I = P/V$, the current in each branch I is
I = Power of each bulb/ potential difference across each bulb
= 10 W/220 V = (1/22) A.
Total current in parallel combination of n branches is
$n \times I$ = Maximum allowable current = 5 A
or n = 5A/ I = 5A/ ((1/22) A) = 110.
An alternative method is as follows:
The resistance of one bulb $r = V^2/P = (220)^2/10 = 4840$ Ω.
Equivalent resistance R of n bulbs joined in parallel is given by
$1/R$ = 1/r+ 1/r + ... n times = n/r
or R = r/n.
Total resistance of parallel combination of n bulbs = r/n. We also have resistance of the circuit = supply voltage/maximum allowable current = 220V/5A = 44 Ω.
So, r/n = 44 Ω, or n = r/44 Ω = 4840 Ω/44 Ω =110.

94. What is the commercial unit of electrical energy? Represent it in terms of joules.

Answer
The commercial unit of electric energy is known as kilowatt hour (1 kW h).
1kW h = 1 kW × 1 hour = 1000 watt × 3600 second =3.6 × 10^6 watt sec.
But 1 watt sec = 1 joule (J)
So, 1 kW h = 3.6 × 10^6 J.

95. An electric geyser has the rating 220V, 2000 W marked on it. What should be the minimum rating, in whole number, of fuse wire that may be required for safe use with this geyser?

Answer
The current through the geyser is $I = P/V$ = 2000 W/220 V = 9.1 A.

96. An electric motor takes 5 A from a 220 V line. Determine the power of the motor and the energy consumed in 2 h.

Answer
The power of motor is
$P = VI = (220\text{ V}) \times (5\text{ A}) = 1100\text{ W}$.
Energy consumed in 2 h $= Pt = (1100\text{ W}) \times (2\text{ h})$
$= 2200$ W h $= 2.20$ kW h.
We can express the result in joules by noting that
1 kW h $= 3.6 \times 10^6$ J,
Energy consumed $= (2.20$ kW h$) \times (3.6 \times 10^6$ J$) = 7.92 \times 10^6$ J.

97. A bulb is rated 240 V, 100 W. Calculate its resistance. Five such bulbs burn for 4 hours daily. Calculate the units of electric energy are consumed per day. What would be the cost of using these bulbs per day at the rate of Rs. 4.0 per unit?

Answer
Power, $P = 100$ W, $V = 200$ V.
Let R be the resistance of the bulb, we have $P = V^2/R$, which gives
$R = V^2/P$
$= (200\text{ V})^2 /100\text{ W} = 400\ \Omega$.
The electric energy consumed /day for running 5 bulbs
$= 100$ W $\times 4$ h $\times 5 = 2000$ W h $= 2$ kWh $= 2$ units.
So, the cost of using these bulbs per day $= 2 \times$ Rs. 4.0 $=$ Rs. 8.0.

98. Compare the power used in the 4 Ω resistor in each of the following circuits: (i) a 6 V battery in series with 2 Ω and 4 Ω resistors, and (ii) a 6 V battery in parallel with 10 Ω and 4 Ω resistors.

Answer
(i) The current through each of the resistances in series
= current I in the series circuit
$= V/(2\ \Omega + 4\ \Omega) = 6$ V/6 Ω $= 1.0$ A.
So, the power P consumed by 4 Ω resistor is
$P = I^2 R = (1.0\text{ A})^2 \times 4\ \Omega = 4.0$ W.
(ii) Potential difference across each of the resistances in parallel = battery voltage. So, the power P consumed by 4 Ω resistor is
$P' = V^2/R = (6\text{ V})^2 / 4\ \Omega = 9$ W.
Hence, P is less than P' i.e., power used by the 4 Ω resistor is lesser in series circuit than in parallel circuit.

99. Three incandescent bulbs of 100 W each are connected in series in an electric circuit. In another circuit another set of three bulbs of the same wattage are connected in parallel to the same source.
(a) Will the bulb in the two circuits glow with the same brightness? Justify your answer.
(b) Now let one bulb in both the circuits get fused. Will the rest of the bulbs continue to glow in each circuit? Give reason.

Answer
(a) Suppose V is the voltage of the source and power $P = 100$ W. Since resistance $= V^2/P$, same wattage of bulbs means that they have equal resistance. Let R be the resistance of each of the bulbs. The total resistance of the bulbs in series $= 3R$. The current through each bulb in series $I_s = V/(3R)$.
When three bulbs of equal resistance R are joined in parallel to the same source, Ohm's law gives
$V = R \times I_p$, where I_p is the equal current flowing in each bulb.
So, $I_p = V/R$. Since, $I_p = 3 I_s$, the parallel combination bulbs will glow more brightly.

(b) When one bulb in series circuit gets fused, the circuit is broken and current becomes zero. So, all the bulbs will stop glowing.
However, in a parallel combination, the current flows simultaneously in different branches. So stopping of current in one branch due to a fused bulb does not result in stopping of current in other branches. As a result, the other bulbs in parallel combination shall continue to glow.

100. If the current *I* through a resistor is increased by 100% (assume that temperature remains unchanged), the increase in power dissipated will be
(a) 100 %
(b) 200 %
(c) 300 %
(d) 400 %

Answer
(c)
(**Hint:** A 100% increase means that the current *I* increases to 2 *I*. Initial power *P* is given as $P = I^2 R$, the new value is $P' = (2I)^2 R = 4 P$. So, the percentage increase = $(4P - P)/P \times 100 = 300\%$.)

101. In an electrical circuit three incandescent bulbs A, B and C of rating 40 W, 60 W and 100 W respectively are connected in parallel to an electric source. Which of the following is likely to happen regarding their brightness?
(a) Brightness of all the bulbs will be the same
(b) Brightness of bulb A will be the maximum
(c) Brightness of bulb B will be more than that of A
(d) Brightness of bulb C will be less than that of B

Answer
(c)
(**Hint:** In parallel combination, a common potential difference *V* exists across each of the bulbs. Now, power $P = V^2/R$ or $R = V^2/P$, which means that resistance of a bulb is inversely proportional to its power rating. So, 60 W bulb B has lower value of resistance than A. Also, power consumed by a bulb = V^2/R. Hence, bulb B consumes more power and will glow brighter than A.)

102. Three 2 Ω resistors, A, B and C, are connected as shown in the diagram given below. Each of them dissipates energy and can withstand a maximum power of 18W without melting. Find the maximum current that can flow through the three resistors.

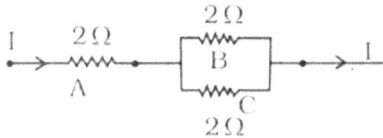

Answer
We know that maximum power *P*, when maximum current *I* flows through a resistance *R* is
$P = I^2 R$, which gives
$I^2 = P/R$
Or, $I = \sqrt{(P/R)}$
Here, $P = 18$ W, $R = 2$ Ω,
So, $I = \sqrt{(18 \text{ W} / (2 \text{ Ω}))} = 3$ A.
Maximum current through resistor A = 3 A. The same current passes through a parallel combination of B and C. Since the resistance values of B and C are equal (= 2 Ω), the current of 3 A is divided equally between B and C. Thus the maximum current through resistors B and C each = (3/2) = 1.5 A.

103. Compare the power used in the 2 Ω resistor in each of the following circuits: (i) a 6 V battery in series with 1 Ω and 2 Ω resistors, and (ii) a 4 V battery in parallel with 12 Ω and 2 Ω resistors.

Answer

(i) The current through each of the resistances in series
= current I in the series circuit
= $V/(1\ \Omega + 2\ \Omega) = 6\ V/3\ \Omega = 2\ A$.
So, the power P consumed by 2 Ω resistor is
$P = I^2 R = (2\ A)^2 \times 2\ \Omega = 8\ W$.
(ii) Potential difference across each of the resistances in parallel = battery voltage. So, the power P consumed by 2 Ω resistor is
$P' = V^2/R = (4\ V)^2/2\ \Omega = 8\ W$.
Hence, $P = P'$ i.e., power used in the 2 Ω resistor in the two circuits is same.

104. Two lamps, one rated 100 W at 220 V and the other 60 W at 220 V, are connected in parallel to electric mains supply. What current is drawn from the line if the supply voltage is 220 V?

Answer

In parallel combination, potential difference across each lamp = supply voltage $V = 220$ V.
Using $I = P/V$, we have current drawn by first lamp $I_1 = P_1/V = 100\ W/220\ V = 0.46\ A$.
Current drawn by second lamp $I_2 = P_2/V = 60\ W/220\ V = 0.27\ A$.
So, current drawn from the line = $I_1 + I_2 = 0.46\ A + 0.27\ A = 0.73\ A$.
(**Hint:** An alternative method is as follows.
Resistance of the first lamp $R_1 = V^2/P_1 = (220\ V)^2/100\ W = 484\ \Omega$
Resistance of the second lamp $R_2 = V^2/P_2 = (220\ V)^2/60\ W = 806.7\ \Omega$.
Since the lamps are in parallel, their equivalent resistance is
$R_p = R_1 R_2/(R_1 + R_2) = 484 \times 806.7/(484 + 806.7) = 302.6\ \Omega$
So, current drawn from the line = $V/R_p = 220\ V/(302.6\ \Omega) = 0.73\ A$.)

105. An electric bulb is rated 60 W, 240 V. Calculate its resistance. If the voltage drops to 192 V, calculate the power consumed and the current drawn by the bulb.

Answer

Power, $P = 60$ W, $V = 240$ V.
Let R be the resistance of the heater coil, we have $P = V^2/R$, which gives
$R = V^2/P$
= $(240\ V)^2/60\ W = 960\ \Omega$.
Resistance remains unchanged when the voltage changes to $V' = 192$ V. So the power consumed will be
$P = (V')^2/R = (192)^2/960\ \Omega = 38.4\ W$.
The current through the coil, $I = V'/R = 192\ V/960\ \Omega = 0.2\ A$.

106. An electric heater of resistance 8 Ω draws 15 A from the service mains for 2 hours. Calculate the rate at which heat is developed in the heater.

Answer

The rate at which heat is developed i.e. power P is given as
$P = I^2 R = (15\ A)^2 \times (8\ \Omega) = 1800\ W$.

107. (a) Two identical resistors each of resistance 10 ohm are connected:
(i) in series
(ii) in parallel, in turn to a battery of 6 V. Calculate the ratio of power consumed in the combination of resistors in the two cases.
(b) Establish the relationship between 1 kWh and SI unit of energy.

Answer

(a) Supply voltage $V = 6$ V.

Resistance $R = 10$
When the resistors are joined in series, the equivalent resistance is $R_s = R + R = 2R = 20\ \Omega$.
The power consumed by the series combination is $P_1 = V^2/R_s = V^2/2R = 36/20 = (9/5)$ W.
When they are joined in parallel, the equivalent resistance is $R_p = R \times R/(R + R) = R/2 = 5\ \Omega$.
The power consumed by the parallel combination is $P_2 = V^2/R_p = V^2/(R/2) = 36/5 = (36/5)$ W.
So, the ratio $P_1/P_2 = (9/5)/(36/5) = 1/4$.
or $P_1: P_2 :: 1:4$.
(b) One kilowatt hour is the energy consumed when 1 kilowatt of power is used for one hour.
So, 1kW h
= 1 kW × 1 hour
= 1000 watt × 3600 second
= 3.6×10^6 watt sec.
But, 1 watt sec = 1 joule (J)
So, 1 kW h = 3.6×10^6 J

108. An electric bulb is rated 220 V and 100 W. When it is operated on 110 V, the power consumed will be
(a) 100 W
(b) 75 W
(c) 50 W
(d) 25 W

Answer
(d)
(**Hint:** First find the resistance R of the bulb. $P = V^2/R$ gives $R = V^2/P = (220)2/100 = 484\ \Omega$. New value of power $P' = (V')^2/R = (110)^2/484 = 25$ W.)

109. Unit of electric power may also be expressed as
(a) volt ampere
(b) kilowatt hour
(c) watt second
(d) joule second

Answer
(a)
(**Hint:** Since electric power $P = V I$, the SI unit of electric power (watt) is given as 1 watt = 1 volt × 1 ampere.)

110. An electric kettle consumes 1 kW of electric power when operated at 220 V. A fuse wire of what rating must be used for it?
(a) 1 A
(b) 2 A
(c) 4 A
(d) 5 A

Answer
(d)
(**Hint:** The current I flowing through the kettle, when used across a line voltage of 220 V is $I = P/V = 1\text{kW}/220\text{ V} = 1000\text{ W}/220\text{ V} = 4.55$ A. For a safe functioning of this appliance, a fuse of 5 A rating should be put in series with the appliance.)

111. What determines the rate at which energy is delivered by a current?

Answer
It is the value of electric power of the source of the current, which determines the rate at which energy is delivered by a current.

114